Multi-Cloud Kubernetes

Designing and implementing multi-cloud Kubernetes architectures

Adam Robertson

www.bpbonline.com

First Edition 2024
Copyright © BPB Publications, India
ISBN: 978-93-55517-029

All Rights Reserved. No part of this publication may be reproduced, distributed or transmitted in any form or by any means or stored in a database or retrieval system, without the prior written permission of the publisher with the exception to the program listings which may be entered, stored and executed in a computer system, but they can not be reproduced by the means of publication, photocopy, recording, or by any electronic and mechanical means.

LIMITS OF LIABILITY AND DISCLAIMER OF WARRANTY

The information contained in this book is true to correct and the best of author's and publisher's knowledge. The author has made every effort to ensure the accuracy of these publications, but publisher cannot be held responsible for any loss or damage arising from any information in this book.

All trademarks referred to in the book are acknowledged as properties of their respective owners but BPB Publications cannot guarantee the accuracy of this information.

Distributors:

BPB PUBLICATIONS
20, Ansari Road, Darya Ganj
New Delhi-110002
Ph: 23254990/23254991

DECCAN AGENCIES
4-3-329, Bank Street,
Hyderabad-500195
Ph: 24756967/24756400

MICRO MEDIA
Shop No. 5, Mahendra Chambers,
150 DN Rd. Next to Capital Cinema,
V.T. (C.S.T.) Station, MUMBAI-400 001
Ph: 22078296/22078297

To View Complete
BPB Publications Catalogue
Scan the QR Code:

Published by Manish Jain for BPB Publications, 20 Ansari Road, Darya Ganj, New Delhi-110002 and Printed by him at Manipal Technologies, Manipal

www.bpbonline.com

Dedicated to

*My parents, **Jeanne** and **Frank**, and
my sisters **Chelsea** and **Jessica***

About the Author

Adam Robertson's early enthusiasm for technology was ignited by the wonder of computers during his infancy. His passion for computers motivated him to earn a Bachelor of Science in Computer Science at the University of Texas, Austin, which became the basis for his creative software development methods which eventually led him to the formation of a new discipline of Software Development, DevOps.

Adam was already embodying the ideals of DevOps before the name was coined. He promoted collaboration and automation, leading initiatives that optimized workflows and encouraged a culture of ongoing enhancement. Today, he is recognized as a prominent figure in this evolving industry, imparting his knowledge as the Head of DevOps at WitnessAI, an AI Security startup.

Adam's competence goes well beyond technical skills. He is an adept team leader, using his vast experience as a DevOps mentor and coach to guide others. He received the esteemed "DevOps Rising Star" award at the 2019 DevOps World Conference in recognition of his significant achievements.

After spending 15 years in the fast-paced environment of Silicon Valley startups, Adam now enjoys the outdoors in Boulder, Colorado. He is enthusiastic about sharing information and participates in speaking events and forums to promote the benefits of the DevOps culture and empower people with the technology he has learned.

About the Reviewer

Werner Dijkerman is a freelance platform, Kubernetes (certified), and Dev(Sec)Ops engineer. He currently focuses on and works with cloud-native solutions and tools, including AWS, Ansible, Kubernetes, and Terraform. He focuses on infrastructure as code and monitoring the correct 'thing' with tools such as Zabbix, Prometheus, and the ELK stack. He has a passion for automating everything and avoiding doing anything that resembles manual work. He is an active reader of comics, self-care/psychology, and IT-related books. He is a also technical reviewer of various books about DevOps, CI/CD, and Kubernetes.

Acknowledgement

First, I would like to thank my family. Throughout the lengthy process of writing this book, their words kept me motivated week after week, chapter after chapter. To my sisters Chelsea and Jessica, who have always kept me grounded and believed in me no matter what, and to my parents Frank and Jeanne, who have always been a guiding light in my life. A special thanks to Aaron Kutzer, who is both an incredible friend and a calming force to help keep me rooted firmly on Earth. To my lifelong friends H Jordan High and Joe "Matt" Hasty, it's been a privilege to share my life's journey with you both.

Second, I want to thank Gil Spencer, a friend and mentor who I turn to for advice and insight in both my personal and professional life.

In addition, I would like to thank my coworkers and colleagues who supported me as we created the new discipline of DevOps. The encouragement and contributions of the DevOps community have been crucial in my professional growth.

In closing, I want to express my gratitude to all of the readers who have shown interest in my book and helped to make it a reality. I look forward to all of the opinions and conversations that will no doubt come in droves.

Preface

The evolving landscape of Cloud Native applications has driven multi-cloud methods to prominence due to the requirements for agility, scalability, and cost-effectiveness. Kubernetes, a powerful container orchestration tool, provides a solution for deploying applications consistently across various cloud environments. Yet, understanding the complexities of multi-cloud Kubernetes can be overwhelming, causing many enterprises to feel uncertain about how to start.

This book serves as a guide to help you realize the capabilities of multi-cloud Kubernetes. We will start by establishing the foundation through project planning and preparation, making sure you have the required knowledge and resources. We will explore the challenges you may face and provide you with solutions to conquer them with confidence.

The book delves into the topic of running Kubernetes on the top 3 cloud providers, AWS, GCP, and Azure, and diverse deployment scenarios, offering practical advice tailored to your requirements. We will explore the intricacies of creating clusters for both stateful and stateless applications, enabling you to make well-informed decisions.

The book progresses with operations, focusing on mastering service mesh for inter-application communication and improved security, as well as adopting GitOps methodologies for automatic and consistent deployments and policy enforcement. We then explore how utilizing proactive monitoring and extensive observability approaches can guarantee the smooth operation of your multi-cloud architecture.

The book ends with a focused case study demonstrating the practical use of what was learned, providing a step-by-step guide on deploying highly available apps on several cloud Kubernetes clusters.

Whether you are an experienced DevOps engineer, cloud architect, or a novice interested in multi-cloud options, this book provides important insights and practical advice. Join us on this journey and harness the potential of multi-cloud Kubernetes for your enterprise.

Chapter 1: Getting Started with Project Planning and Preparation – This chapter guides you through planning your multi-cloud Kubernetes project, covering prerequisites, essential tools, and key considerations for a smooth implementation.

Chapter 2: Running Multi-cloud Kubernetes – This chapter talks about the unique challenges of managing Kubernetes across different cloud providers, including network connectivity, security considerations, and cost optimization strategies.

Chapter 3: Scenarios of Multi-Cloud Deployment – This chapter explores the various deployment scenarios like disaster recovery, workload migration, and hybrid cloud strategies, understanding their benefits and potential pitfalls.

Chapter 4: Stateful Application Kubernetes Clusters Design – This chapter will enable you to learn the art of designing resilient Kubernetes clusters for stateful applications, ensuring data persistence and availability across multiple cloud environments.

Chapter 5: Stateless Application Kubernetes Clusters Design – Optimize your infrastructure for stateless applications with efficient cluster design patterns and scaling strategies tailored to multi-cloud deployments.

Chapter 6: Service Mesh: Operations – This chapter will help you to master the power of service meshes for seamless service-to-service communication and discover operational best practices for managing them across your multi-cloud landscape.

Chapter 7: Service Mesh: Security – This chapter aims to elevate your security posture with service meshes, learning how to implement access control, encryption, and monitoring strategies for enhanced protection.

Chapter 8: GitOps Method of Workload Deployment – This chapter will help you embrace the GitOps approach for automated and consistent workload deployments, ensuring configuration drift is a thing of the past.

Chapter 9: GitOps Method of Policy Deployment – After reading this chapter, you will be able to simplify policy management with GitOps, ensuring governance and compliance across your multi-cloud Kubernetes deployments.

Chapter 10: Proactive Monitoring of the Clusters – This chapter will enable you to stay ahead of the issues with proactive monitoring techniques, learning how to collect and analyze metrics for optimal performance and troubleshooting.

Chapter 11: Enabling Comprehensive Observability – This chapter will help you to gain deep insights into your multi-cloud infrastructure with comprehensive observability practices, empowering data-driven decision making.

Chapter 12: Securing your Clusters – This chapter will enable you to prioritize security from the ground up. It delves into the robust security practices for securing your multi-cloud Kubernetes clusters, protecting your applications and data.

Chapter 13: Case Study: Deploying Highly Available Application on Multiple-Cloud Kubernetes – The case studies in this chapter will walk you through a real-world deployment of highly available applications across multiple cloud Kubernetes clusters, showcasing the practical implementation of best practices.

Code Bundle and Coloured Images

Please follow the link to download the
Code Bundle and the *Coloured Images* of the book:

https://rebrand.ly/0603wgp

The code bundle for the book is also hosted on GitHub at

https://github.com/bpbpublications/Multi-Cloud-Kubernetes

In case there's an update to the code, it will be updated on the existing GitHub repository.

We have code bundles from our rich catalogue of books and videos available at **https://github.com/bpbpublications**. Check them out!

Errata

We take immense pride in our work at BPB Publications and follow best practices to ensure the accuracy of our content to provide with an indulging reading experience to our subscribers. Our readers are our mirrors, and we use their inputs to reflect and improve upon human errors, if any, that may have occurred during the publishing processes involved. To let us maintain the quality and help us reach out to any readers who might be having difficulties due to any unforeseen errors, please write to us at :

errata@bpbonline.com

Your support, suggestions and feedbacks are highly appreciated by the BPB Publications' Family.

Did you know that BPB offers eBook versions of every book published, with PDF and ePub files available? You can upgrade to the eBook version at www.bpbonline.com and as a print book customer, you are entitled to a discount on the eBook copy. Get in touch with us at :

business@bpbonline.com for more details.

At **www.bpbonline.com**, you can also read a collection of free technical articles, sign up for a range of free newsletters, and receive exclusive discounts and offers on BPB books and eBooks.

Piracy

If you come across any illegal copies of our works in any form on the internet, we would be grateful if you would provide us with the location address or website name. Please contact us at **business@bpbonline.com** with a link to the material.

If you are interested in becoming an author

If there is a topic that you have expertise in, and you are interested in either writing or contributing to a book, please visit **www.bpbonline.com**. We have worked with thousands of developers and tech professionals, just like you, to help them share their insights with the global tech community. You can make a general application, apply for a specific hot topic that we are recruiting an author for, or submit your own idea.

Reviews

Please leave a review. Once you have read and used this book, why not leave a review on the site that you purchased it from? Potential readers can then see and use your unbiased opinion to make purchase decisions. We at BPB can understand what you think about our products, and our authors can see your feedback on their book. Thank you!

For more information about BPB, please visit **www.bpbonline.com**.

Join our book's Discord space

Join the book's Discord Workspace for Latest updates, Offers, Tech happenings around the world, New Release and Sessions with the Authors:

https://discord.bpbonline.com

Table of Contents

1. **Getting Started with Project Planning and Preparation** 1
   Introduction 1
   Structure 1
   Objectives 2
   The rise of containerization and Kubernetes 2
   Need for multi-cloud Kubernetes management 2
       Defining your objectives 2
       Defining your requirements 3
       Defining your scope 5
       Assess your existing infrastructure 6
       Choosing the right cloud providers 7
       Developing a multi-cloud strategy 8
   Addressing network requirements 9
       Intra-cluster networking 9
       Inter-cluster networking 9
       Load balancing 9
       DNS configuration 10
       IP address management 10
       Network security 10
       Network monitoring and visibility 10
       Network resilience and high availability 10
       Planning for security and compliance 10
   Designing the Kubernetes cluster 12
       Define cluster objectives 12
       Establish monitoring and logging 13
       Defining backup and disaster recovery strategies 15
       Set up CI/CD pipelines 16
       Investing in training and skill development 17
       Setting up your local development environment 18
   Conclusion 19

2. Running Multi-Cloud Kubernetes 21
Introduction 21
Structure 21
Objectives 22
Rationale behind multi-cloud Kubernetes 22
Evolution of multi-cloud Kubernetes 23
    Pre-Kubernetes era 24
    Kubernetes emergence (2014) 24
    Managed Kubernetes services (2017 onward) 24
    Multi-cloud Kubernetes tools and solutions 25
    Multi-cloud service mesh and networking solutions 25
    GitOps and Infrastructure as Code 25
Pros of multi-cloud Kubernetes 25
Cons of multi-cloud Kubernetes 27
Multi-cloud Kubernetes considerations 28
Hypothetical use cases 29
    Use case 1: High availability and redundancy 29
    Use case 2: Improved geographical coverage and latency reduction 30
    Use case 3: Cost optimization and flexibility 31
    Use case 4: Disaster recovery and business continuity 31
    Use case 5: Compliance and regulatory requirements 32
Hypothetical case study 32
Industry best practices 34
Conclusion 35

3. Scenarios of Multi-Cloud Deployment 37
Introduction 37
Structure 37
Objectives 38
Stateful and stateless applications 38
    Stateful applications 38
    Stateless applications 38
Differences between two types of clusters 39
    Data persistence and management 39
    Scalability 39

		High availability and fault tolerance *39*
		Deployment and maintenance complexity.. *40*
	Stateful application-based clusters 41
		Stateful applications *41*
		Kubernetes and stateful applications *42*
		Complexity in multi-cloud Kubernetes deployments *42*
	Use cases where stateful clusters are best suited 43
	Stateless application-based clusters... 44
		Implications *45*
		Complexities.. *45*
	Use cases where stateless clusters are best suited 46
	Use cases 47
	Advantages.. 47
	Challenges 48
	Conclusion... 48

4. **Stateful Application Kubernetes Cluster Design**... **49**
	Introduction.. 49
	Structure.. 49
	Objectives.. 50
	Overview of stateful applications 50
	Considerations for stateful application clusters 51
	Data storage strategies across multiple cloud providers... 52
	Sharing real-time data across clusters in different cloud providers 54
	Best practices for stateful application design 55
	Stateful application Kubernetes cluster design patterns 56
	Example of stateful application deployments... 57
		Challenges of deploying Redis *57*
		Multi-cloud Redis system design *58*
			Architecture overview *58*
			Storage options *59*
			Data replication and synchronization *59*
			Networking... *59*
		Project plan *60*
			Project name *60*

Project scope	60
Objectives	60
Key deliverables	60
Project stakeholders and roles	61
Project timeline	61
Resources	61
Risk management plan	62
Architectural diagram	62
Implementation steps for deploying Redis	63
Prerequisites	63
Versions	63
Base configuration	63
Storage class	65
StatefulSet and PVC	65
Create the resources	67
Verify the Redis deployment	68
Test the Redis deployment	68
Expose the Redis database	69
Test out the public endpoint	70
Conclusion	71

5. Stateless Application Kubernetes Cluster Design 73

Introduction	73
Structure	73
Objectives	74
Overview of stateless applications	74
Architecture and design considerations	76
Designing for scalability	76
Data persistence and management	76
Load balancing and networking	76
Fault tolerance and health checks	76
Containerization and image design	76
Configuration management	77
Interoperability and portability	77
Kubernetes resources	77

Network	78
Kubernetes resources for stateless applications	80
Load balancing and scaling stateless applications	81
Load balancing	81
Pod scaling	81
Horizontal Pod Autoscaling	82
Cluster Autoscaling	82
Auto healing	82
Stateless application deployment patterns	82
Single cloud deployment pattern	83
Blue-green deployment	83
Canary deployment	83
Multi-cloud deployment pattern	83
Hands-on system design and project plan	84
Brief introduction to the example application	85
Challenges in deploying applications	86
Why this application is representative of common stateless applications	87
Multi-cloud NGINX web server system design	87
Architecture overview	87
Network design	88
Load balancing	88
Scaling	89
Project plan for deploying NGINX web server on EKS	89
Project plan	90
Project scope	90
Objectives	90
Key deliverables	90
Project stakeholders and roles	90
Project timeline	91
Resources	91
Risk management plan	91
Architectural diagram	92
Implementation steps for deploying NGINX web server	93
Prerequisites	93

Versions	93
Base configuration	94
Deployment, ReplicaSet and Pods	96
Kubernetes ReplicaSet	96
Kubernetes Pod	96
Create the resources	97
Verify the NGINX web server deployment	98
Test the NGINX web server deployment	99
Expose the NGINX web server	99
Test out the public endpoint	100
Conclusion	101

6. Service Mesh: Operations 103

Introduction	103
Structure	104
Objectives	104
Introducing service mesh and its operational benefits	104
History of service mesh	106
Common use cases for service mesh	107
Deciding when to implement a service mesh	108
Understanding service mesh components	109
Control plane	109
Data plane	110
Deciding which control plane manager to use	110
Data plane vendor comparison matrix	111
Deciding which data plane manager to use	112
Data plane vendor comparison matrix	113
Overview of Istio and Envoy	114
Istio as a control plane	114
Pros of Istio	114
Cons of Istio	115
Envoy as a data plane	115
Pros of Envoy	115
Cons of Envoy	115
How do they work together	116

Service mesh traffic control 116
    Routing and load balancing 117
        Routing 117
        Load balancing 117
    Service discovery 118
    Fault injection and traffic mirroring 119
        Fault injection 119
        Traffic mirroring 119
    Rate limiting and quota management 120
        Rate limiting 120
        Quota management 120
    Access control and policy enforcement 121
    TLS and mTLS 122
Implementing resiliency 122
    Retries 123
    Timeouts 124
    Circuit breakers 124
    Health checks 125
    Load balancing 126
    Rate limiting 127
    Outlier detection 128
Observability in service mesh 128
Conclusion 129

7. **Service Mesh: Security** 131
    Introduction 131
    Structure 131
    Objectives 132
    Introduction to service mesh security 132
    Security principles in a service mesh 133
        Zero-trust network 133
        Least privilege 135
        Mutual authentication 136
        Secure communication 136
        Fine-grained access control 137

Policy enforcement	138
Observability	139
Secure ingress/egress	140
Auditability	141
Traffic encryption with mutual TLS	142
Mutual TLS	142
Implementing mTLS in a service mesh	142
Authorization and access control	143
Authorization	143
Access control	143
Network policy and isolation	144
Network policy	144
Isolation	144
Security policies in a control plane	145
Authorization policies	146
Audit logging policies	147
Network policies	148
Peer authentication policies	148
Threat modelling and security best practices	149
Security considerations	150
Conclusion	151
8. GitOps Method of Workload Deployment	**153**
Introduction	153
Structure	153
Objectives	154
Introduction to GitOps	154
Core principles of GitOps	155
Declarative infrastructure in GitOps	155
Version control as single source of truth	156
Automated synchronization	157
Immutable infrastructure and image-based deployments	158
Operational procedures through Pull Requests	159
Wrapping up GitOps principals	160
Benefits and challenges of GitOps	160

- Benefits of GitOps *161*
- Challenges of GitOps *161*
- GitOps workflow 162
- GitOps tools 163
- Implementing GitOps in a service mesh 164
  - Single cluster *164*
  - Multi-region clusters *165*
  - Multi-cloud Kubernetes clusters *165*
- Securing GitOps workflow 165
- GitOps best practices 166
- Conclusion 167

9. GitOps Method of Policy Deployment 169
- Introduction 169
- Structure 169
- Objectives 170
- Introduction to GitOps policy deployment 170
- Policy as code 171
- Core principles of GitOps for policy deployment 171
  - Policy as code *172*
  - Automated policy enforcement *172*
  - Continuous monitoring and reconciliation *172*
  - Automated validation and testing *172*
  - Observability and auditability *173*
  - Security and compliance *173*
- Implementing GitOps for policy deployment 173
- Tools for GitOps policy deployment 174
- Securing GitOps policy deployment 175
- GitOps policy deployment best practices 176
  - Clearly define policies *177*
  - Automate everything *178*
  - Version control *179*
  - Regularly review and update policies *180*
  - Testing and validation *180*
  - Observability *181*

  Plan for failure 182
 Conclusion 184

10. Proactive Monitoring of the Clusters 185
 Introduction 185
 Structure 185
 Objectives 186
 Introduction to proactive monitoring 186
 Importance and benefits of proactive monitoring 187
 Challenges in proactive monitoring 188
 Tools and techniques for proactive monitoring in Kubernetes 189
 Implementing proactive monitoring 190
  Understanding your system 190
   Problem it solves 191
   Solving the problem 191
  Selecting the right tools 191
   Problem it solves 192
   Solving the problem 192
  Configuring the tools 193
   Problem it solves 193
   Solving the problem 193
  Building dashboards 194
   Problem it solves 194
   Solving the problem 194
  Setting up alerts 195
   Problem it solves 195
   Solving the problem 195
  Testing your setup 196
   Problem it solves 196
   Solving the problem 196
  Iterative improvement 197
   Problem it solves 197
   Solving the problem 197
 Understanding and using metrics, logs, traces and alerts 198
  Metrics 199

    Traces *199*

    Logs *200*

    Alerts *200*

  Best practices for proactive monitoring 200

  Securing your monitoring stack 201

  Proactive monitoring in service mesh 202

  Conclusion 203

11. Enabling Comprehensive Observability 205

  Introduction 205

  Structure 205

  Objectives 206

  Introduction to comprehensive observability 206

  Importance of observability in multi-cloud Kubernetes 207

    Complexity and variability *207*

    Troubleshooting and root cause analysis *208*

    Performance optimization *208*

    Proactive problem detection *208*

    Insight and understanding *208*

    Cost management *209*

  Understanding the three pillars of observability 210

    Logs *210*

    Metrics *210*

    Traces *210*

  Tools and techniques for achieving comprehensive observability 211

    Prometheus *211*

    Grafana *211*

    Elastic Stack *212*

    Jaeger and Zipkin *212*

    Fluentd/Fluent Bit *212*

    Service mesh *213*

    OpenTelemetry *213*

  Implementing observability 213

  Observability with service mesh 215

  Best practices for comprehensive observability 216

   Use standardized and structured logging 216
   Instrumentation 216
   Leverage distributed tracing 216
   Set up efficient alerting 217
   Correlation of data 217
   Embrace service mesh 218
   Continual review and improvement 218
   Use open standards and open-source software 218
   Ensure security 218
  Challenges in implementing observability 219
   Volume of data 219
   Diversity of data sources 219
   Complexity of systems 219
   Alert fatigue 219
   Security and compliance 220
   Skills and knowledge gap 220
   Cost 220
  Case studies: Observability in action 220
   E-commerce giant embraces observability for Black Friday 220
   Streaming service avoids outage during global event 221
  Conclusion 221

12. Securing Your Clusters **223**
  Introduction 223
  Structure 223
  Objectives 224
  Introduction to Kubernetes security 224
  Four Cs of cloud-native security 225
   Code 225
   Container 227
   Cluster 228
   Cloud 229
  Securing your infrastructure: Best practices 230
  Network security in Kubernetes 231
  Securing externally generated client requests 232

Securing intra-cluster app-to-app communication 233
Tools and techniques for Kubernetes security 234
Case studies: Security practices in action.. 235
    Case study 1: FinTech company implements zero trust networking 235
    Case study 2: E-commerce firm secures service mesh with envoy 235
    Case study 3: Healthcare provider secures patient data 235
    Case study 4: Media company handles high traffic securely 236
    Case study 5: Software company streamlines compliance 236
Best practices for implementing security 236
Challenges and solutions in Kubernetes security 237
Security in the context of service mesh 239
Conclusion 239

13. Case Study: Deploying Highly Available Application on Multi-Cloud Kubernetes 241

Introduction 241
Structure 241
Objectives 242
Introduction to HA applications 242
Definition and importance of HA 242
Distinction between availability, reliability, and scalability 243
Unlocking HA with Kubernetes 243
    Service mesh with Istio and Envoy 244
    Kubernetes and multi-cloud 244
Designing for HA 245
    Factors influencing design 245
        Latency 245
        Redundancy 245
        Replication 245
    Stateful vs. stateless applications 246
        Stateless applications 246
        Stateful applications 246
    Role of persistent storage and network design 246
        Persistent storage 246
        Network design 247

- Deploying applications on AWS EKS 247
  - Key features and benefits 247
    - Key features 247
    - Benefits 248
  - Steps for setting up an EKS cluster 248
    - EKS control plane creation 248
    - Node setup 248
  - Deploying an application using Helm and other EKS tools 249
    - Setting up Helm 249
    - Deploying using Helm 249
    - EKS tools and utilities 249
  - EKS with Istio and Envoy 249
- Deploying applications on GCP GKE 250
  - Advantages 250
  - Configuring a GKE cluster: Step-by-step guide 250
    - Initial setup 250
    - Cluster creation 251
  - Application deployment using GCP-native tools 251
    - Container registry 251
    - Deploying with cloud build and cloud source repositories 251
    - Cloud deployment manager 252
  - GKE with Istio and Envoy 252
- Implementing cross-cloud communication 252
  - Exploring challenges of cross-cloud communication 252
  - Configuring secure networking between AWS and GCP 253
  - Demonstrating communication between applications across clouds 253
    - Service entry and egress gateways 253
    - Demonstration setup 254
- Monitoring and managing HA 254
  - Importance of proactive monitoring in HA applications 254
  - Tools: Prometheus, Grafana, and cloud-native solutions 255
  - Alerting, Logging, and Disaster recovery strategies 255
    - Alerting 255
    - Logging 256

  Disaster recovery strategies... *256*
Best practices for HA application deployment 256
  Ensuring redundancy at every layer *257*
  Automated scaling *257*
  Regular backup, testing, and update policies *258*
Challenges and considerations.. 258
  Data sovereignty and regulatory *259*
  Cost management and optimization strategies... *259*
  Vendor lock-in and interoperability challenges *259*
Case study: Deploying a multi-cloud HA application... 260
  Project overview... *260*
  Milestones... *260*
    Milestone 1: Planning and architecture... *261*
    Milestone 2: Infrastructure setup *261*
    Milestone 3: Application deployment *261*
    Milestone 4: HA and failover *261*
    Milestone 5: Security and observability... *261*
    Milestone 6: Continuous deployment *262*
    Milestone 7: Testing and validation *262*
    Milestone 8: Performance optimization *262*
    Milestone 9: Monitoring and alerting *262*
    Milestone 10: Documentation and knowledge sharing.......................... *262*
Conclusion... 262

Index..**263-270**

Chapter 1
Getting Started with Project Planning and Preparation

Introduction

In today's rapidly evolving cloud landscape, architecting a multi-cloud Kubernetes environment has become an increasingly important aspect of modern application development and deployment. This chapter will cover the planning and preparation crucial to ensure the seamless integration, scalability, and resilience of your Kubernetes clusters across multiple cloud providers.

This chapter will provide you with a comprehensive introduction to the fundamental concepts, methodologies, and best practices to kick off a successful multi-cloud Kubernetes project, allowing you to harness the full potential of diverse cloud infrastructures while maintaining consistent and efficient application management across various platforms. While we try to be vendor agnostic, we will suggest popular tools and products that are used in the industry.

Structure

The chapter covers the following topics:

- The rise of containerization and Kubernetes
- The need for multi-cloud Kubernetes management

- Addressing network requirement
- Designing the Kubernetes cluster

Objectives

By the end of this chapter, you will be in a good position to start building out your multi-cloud Kubernetes cluster having ensured a successful, efficient, and cost-effective deployment of your applications across multiple cloud providers.

The rise of containerization and Kubernetes

Over the past decade, containerization has become a key technology in the world of software development and deployment. Containers package applications and their dependencies into a single, portable unit, providing an isolated environment for running applications consistently across various platforms. Docker, one of the most popular containerization platforms, has played a significant role in popularizing container technology.

Kubernetes, an open-source container orchestration platform, was developed by Google to manage the deployment, scaling, and operation of containerized applications. Kubernetes has rapidly gained adoption due to its flexibility, scalability, and strong ecosystem. It enables organizations to manage containers at scale, providing features such as automated rollouts and rollbacks, self-healing, load balancing, and more.

Need for multi-cloud Kubernetes management

As organizations continue to embrace the cloud, many are adopting multi-cloud strategies to avoid vendor lock in, optimize costs, and improve application performance and resilience. A multi-cloud approach enables businesses to distribute their workloads across multiple cloud providers and leverage the best services and features from each. In this context, Kubernetes has emerged as a powerful tool for managing containerized applications across multiple clouds.

Building a multi-cloud Kubernetes environment requires careful planning, expertise in Kubernetes and cloud services, and a solid understanding of networking, security, and other critical aspects. This chapter will provide an in-depth look at the prerequisites needed before starting to build a multi-cloud Kubernetes environment.

Defining your objectives

Before starting the process of building a multi-cloud Kubernetes environment, it is crucial to define your organization's objectives for adopting this approach. These objectives will help guide your decision-making process and inform your strategy as you build out your multi-cloud environment. While the objectives will vary from company to company, some common objectives for adopting a multi-cloud Kubernetes environment include the following:

- **Cost optimization**: By distributing workloads across multiple cloud providers, organizations can leverage the most cost-effective solutions for their specific needs. This may include using reserved instances, spot instances, or leveraging cloud provider-specific discounts and incentives.
- **High availability and redundancy**: Multi-cloud environments provide an additional layer of redundancy by ensuring that your applications are not solely reliant on a single cloud provider. This helps minimize the risk of downtime due to provider outages, maintenance events, or other unforeseen issues.
- **Disaster recovery**: A multi-cloud strategy can play a critical role in your disaster recovery plans by allowing you to store backups and replicas of your applications and data across different cloud providers. This ensures that your applications can continue to operate in the event of a significant outage or disaster affecting one of your cloud providers.
- **Compliance and data sovereignty**: Some organizations have regulatory or legal requirements related to where their data is stored and processed. By deploying Kubernetes clusters across multiple cloud providers with data centers in different geographic regions, organizations can ensure compliance with these requirements and avoid potential penalties.
- **Cloud bursting**: Cloud bursting in a multi-cloud Kubernetes environment involves dynamically allocating additional computing resources from a secondary cloud provider to handle sudden increases in demand, enabling scalable, cost-effective, and resilient application deployment that seamlessly distributes workloads between primary and bursting cloud providers.
- **Performance and latency**: Running Kubernetes clusters on multiple cloud providers can help optimize application performance and reduce latency for end-users. By distributing workloads across providers with data centers closer to your users, you can improve response times and overall user experience.

Defining your requirements

Adopting a multi-cloud Kubernetes approach requires addressing several key requirements to ensure a successful and seamless deployment across multiple cloud providers. It is important to prioritize your requirements using your business needs as a guide. Here are some essential requirements to consider:

- **Choose your operating system (OS)**: In specifying requirements for a multi-cloud Kubernetes environment, the choice of the OS is critical as it influences compatibility, security, and performance. Opting for OSs that are well-supported by

Kubernetes distributions and cloud providers is essential for seamless integration and efficient resource utilization. Compatibility with container runtimes, system libraries, and kernel versions must be carefully assessed to ensure a robust and unified foundation across diverse cloud environments.

- **Choose your Kubernetes distribution**: Similarly, selecting an appropriate Kubernetes distribution is paramount in defining requirements. Different distributions come with varying features, management tools, and support levels. It is crucial to evaluate factors like ease of installation, scalability, security features, and compatibility with cloud-native services. Consistency in the Kubernetes distribution across clouds simplifies management, deployment, and ensures a cohesive experience for developers and operators, fostering a more efficient multi-cloud Kubernetes ecosystem.

- **Clear objectives and strategy**: Define your business goals, technical objectives, and the reasons for adopting a multi-cloud strategy. This can include cost optimization, improved resilience, leveraging provider-specific services, or avoiding vendor lock-in.

- **Expertise and skillset**: Ensure that your team has the necessary skills and expertise to manage Kubernetes clusters across multiple cloud providers. This may involve investing in training, hiring experienced personnel, or collaborating with external partners.

- **Unified management and orchestration**: Adopt tools and platforms that enable consistent management and orchestration of Kubernetes clusters across different cloud providers, such as Kubernetes Federation, Anthos, or Rancher.

- **Infrastructure as Code (IaC)**: Use IaC tools like Terraform, Pulumi, or AWS CloudFormation to automate the provisioning and management of your multi-cloud Kubernetes infrastructure, ensuring consistency, scalability, and version control.

- **Networking and security**: Design and implement a robust networking and security architecture that allows secure communication between clusters and services across multiple cloud providers. This may involve setting up VPC peering, VPN connections, or using multi-cloud service meshes like Istio or AWS App Mesh.

- **Monitoring and observability**: Choose monitoring and observability tools that provide comprehensive visibility into the performance, health, and security of your multi-cloud Kubernetes environment, such as Prometheus, Grafana, or Datadog.

- **Data management and storage**: Implement a data management strategy that ensures consistent and reliable data storage and access across multiple cloud providers. This may involve using cloud-agnostic storage solutions like Rook, OpenEBS, or Portworx.

- **Continuous Integration/Continuous Deployment (CI/CD):** Establish a CI/CD pipeline that can deploy applications across multiple Kubernetes clusters and cloud providers, using tools like Jenkins, GitLab CI, or GitHub Actions.
- **Cost management:** Implement a cost management strategy to monitor, control, and optimize your spending across multiple cloud providers. This can include using cloud provider-specific cost management tools or third-party solutions like CloudHealth or CloudZero.
- **Compliance and governance:** Ensure that your multi-cloud Kubernetes environment adheres to organizational policies, industry standards, and regulatory requirements. This may involve implementing access controls, auditing, logging, and data protection measures.

Addressing these requirements will help you build a robust, efficient, and maintainable multi-cloud Kubernetes environment that meets your organization's objectives and leverages the advantages of multiple cloud providers.

Defining your scope

When adopting a multi-cloud Kubernetes approach, defining the scope is crucial to ensure the project is focused, manageable, and aligned with your organization's goals. It is important to include items in and out of scope along with an additional *Nice to Have* component where non-essential items can be listed. This is extremely important as *scope creep* can prolong the project and add in too many elements to an already complex operation. The scope should encompass the following aspects:

- **Business objectives:** Clearly outline the business goals driving the adoption of a multi-cloud Kubernetes strategy, such as improving application resilience, optimizing costs, or avoiding vendor lock-in.
- **Technical objectives:** Define the technical goals of the multi-cloud Kubernetes environment, including performance, scalability, security, and maintainability.
- **Cloud providers:** Identify the cloud providers you plan to use for your multi-cloud Kubernetes environment, taking into consideration their offerings, pricing, and compliance with your organization's requirements.
- **Target applications:** Determine the applications or services that will be deployed in the multi-cloud Kubernetes environment, considering factors like existing infrastructure, dependencies, and migration requirements.
- **Infrastructure components:** Specify the infrastructure components needed to build the multi-cloud Kubernetes environment, such as compute, storage, and networking resources, as well as any provider-specific services or tools.
- **Management and orchestration:** Define the tools and platforms required for managing and orchestrating Kubernetes clusters across multiple cloud providers, such as Kubernetes Federation, Anthos, or Rancher.

- **Security and compliance**: Establish the security and compliance requirements for your multi-cloud Kubernetes environment, including data protection, access control, auditing, and adherence to industry standards and regulations.
- **Monitoring and observability**: Specify the monitoring and observability tools needed to maintain visibility into the performance, health, and security of your multi-cloud Kubernetes environment.
- **CI/CD pipeline**: Identify the tools and processes required for deploying applications across multiple Kubernetes clusters and cloud providers, such as Jenkins, GitLab CI, or GitHub Actions.
- **Cost management**: Define the cost management strategy to monitor, control, and optimize spending across multiple cloud providers.
- **Timeline and milestones**: Set a realistic timeline with key milestones for implementing your multi-cloud Kubernetes environment, taking into account potential dependencies, risks, and resource constraints.
- **Roles and responsibilities**: Assign clear roles and responsibilities to team members and stakeholders involved in the project, ensuring effective communication and collaboration throughout the process.

By defining the scope of your multi-cloud Kubernetes project, you can establish a clear roadmap for its implementation, allocate resources effectively, and ensure that your team remains focused on achieving the desired business and technical outcomes.

Assess your existing infrastructure

Before embarking on a multi-cloud Kubernetes deployment, it is essential to assess your existing infrastructure to determine its suitability for the new environment. This assessment should include the following:

- **On premises systems**: Evaluate your current on-premises systems, including compute, storage, and networking infrastructure. Determine whether any of these systems need to be updated, replaced, or integrated with your multi-cloud Kubernetes environment.
- **Current cloud provider usage**: Review your organization's existing usage of cloud providers and identify any existing services, applications, or resources that may need to be migrated, integrated, or replicated in your multi-cloud Kubernetes environment.
- **Existing containerized applications**: If your organization is already using containers and Kubernetes, evaluate the existing applications and identify any potential issues or challenges that may arise when transitioning to a multi-cloud environment. This may include adjusting configurations, modifying deployment strategies, or addressing dependencies on cloud provider-specific services.

Choosing the right cloud providers

Choosing the right cloud providers for your multi-cloud Kubernetes environment is critical to ensure smooth deployment and management of containerized applications. Here are some factors to consider when evaluating and selecting cloud providers:

- **Managed Kubernetes services**: Check if the cloud providers offer managed Kubernetes services, such as **Google Kubernetes Engine (GKE)**, **Amazon Elastic Kubernetes Service (EKS)**, or **Azure Kubernetes Service (AKS)**. These services can simplify cluster deployment, management, and scaling, reducing the operational overhead for your team.
- **Integration with other services**: Consider how well the cloud providers integrate with other services that you may require, such as databases, storage, networking, and monitoring. Seamless integration with essential services can improve the efficiency of your multi-cloud Kubernetes environment.
- **Availability and performance**: Analyze the cloud providers' data center locations, network latency, and performance to ensure that they can meet your application's availability and performance requirements. Selecting providers with data centers near your users can help reduce latency and improve user experience.
- **Cost-efficiency**: Compare the pricing models, discounts, and cost structures of the different cloud providers to determine which ones best align with your budget and cost optimization goals. This may include factors such as compute, storage, and data transfer costs, as well as the availability of cost-saving features like autoscaling and spot instances.
- **Compliance and security**: Evaluate the cloud providers' security and compliance features, ensuring that they meet your industry-specific regulations and security requirements. This may include data encryption, access controls, and certifications such as GDPR, HIPAA, or PCI DSS.
- **Scalability and flexibility**: Assess the providers' ability to scale resources to meet the growing demands of your containerized applications. This includes the availability of autoscaling features, support for custom resource configurations, and options for horizontal and vertical scaling.
- **Support and community**: Investigate the level of support, documentation, and community engagement offered by each cloud provider. Strong support and an active community can help you troubleshoot and resolve issues quickly and efficiently.
- **Vendor lock-in risk**: Consider the extent to which each provider's services and technologies might lock you into their ecosystem. Opting for providers with open standards and APIs can help minimize vendor lock-in risk and ease migration between cloud environments.

By carefully evaluating these factors and weighing them against your organization's objectives, you can select the right cloud providers for your multi-cloud Kubernetes environment, ensuring smooth deployment, management, and scaling of containerized applications across different clouds.

Developing a multi-cloud strategy

Developing a successful multi-cloud strategy requires a thorough understanding of your organization's objectives, technical requirements, and resources. Here is a step-by-step guide to help you develop an effective multi-cloud strategy:

1. **Define your objectives:** Begin by identifying the goals you want to achieve with your multi-cloud strategy, such as cost optimization, high availability, disaster recovery, improved performance, or compliance with data sovereignty regulations.

2. **Assess your existing infrastructure:** Evaluate your current infrastructure, including on-premises systems and current cloud provider usage, to understand its suitability for multi-cloud deployment and identify any necessary upgrades or changes.

3. **Choose the right cloud providers:** Research and compare various cloud providers based on factors like managed Kubernetes services, integration with other services, availability and performance, cost-efficiency, compliance and security, scalability and flexibility, and support and community engagement.

4. **Allocate workloads and resources:** Determine how you will distribute your workloads and resources across the selected cloud providers. This may involve prioritizing workloads based on factors like performance requirements, cost, data locality, and regulatory constraints.

5. **Design your multi-cloud architecture:** Develop a high-level architecture for your multi-cloud environment, including Kubernetes cluster design, networking, storage, security, and management. This may involve a federated, independent, or hybrid architecture, depending on your specific requirements.

6. **Plan for networking and connectivity:** Address the networking requirements for your multi-cloud environment, including intra-cluster and inter-cluster networking, load balancing, DNS configuration, and integrating with cloud provider networking services.

7. **Implement security and compliance measures:** Establish security and compliance requirements for your multi-cloud environment, such as authentication and authorization, network security, container security, data encryption, and adherence to industry-specific regulations.

8. **Establish monitoring and logging:** Plan for centralized logging and monitoring solutions that provide visibility into your multi-cloud Kubernetes environment, including cluster health, resource utilization, and application performance.

9. **Define backup and disaster recovery strategies:** Develop a comprehensive backup and disaster recovery plan, including multi-cloud failover, data backup and restore strategies, and regular testing of disaster recovery procedures.

10. **Set up CI/CD pipelines:** Design and implement continuous integration and deployment pipelines that support multi-cloud Kubernetes deployments, including integration with cloud provider services and deployment strategies tailored to multi-cloud environments.
11. **Train and upskill your team:** Invest in training and skill development for your team to ensure they have the necessary expertise in Kubernetes, containerization technologies, and cloud provider-specific services to effectively manage and maintain a multi-cloud Kubernetes environment.

By following these steps and continuously reviewing and adjusting your strategy as needed, you can successfully implement a multi-cloud Kubernetes environment that meets your organization's objectives, maximizes the benefits of using multiple cloud providers, and ensures efficient deployment and management of containerized applications.

Addressing network requirements

Addressing networking requirements in a multi-cloud Kubernetes environment is crucial for seamless communication between applications and services. In the following section, we will discuss the key aspects that should be considered when planning for networking in a multi-cloud setup.

Intra-cluster networking

Ensure that the networking within each Kubernetes cluster is properly configured, including pod-to-pod and pod-to-service communication. Choose an appropriate **Container Network Interface (CNI)** plugin that supports your requirements and is compatible with your cloud providers.

Inter-cluster networking

For communication between different Kubernetes clusters, you need to establish secure and efficient inter-cluster networking. This can be achieved using VPNs, dedicated interconnects, or other secure communication methods provided by your cloud providers. Additionally, you may consider service mesh solutions like **Istio** or **Linkerd** to manage inter-cluster communication and policies.

Load balancing

Implement load balancing for distributing incoming traffic across your services and ensuring high availability. This may involve using cloud provider-managed load balancers, Kubernetes Ingress controllers, or third-party load balancing solutions.

DNS configuration

Properly configure DNS for service discovery in your multi-cloud Kubernetes environment. This can include using Kubernetes-native DNS solutions like **CoreDNS** or integrating with external DNS services provided by your cloud providers.

IP address management

Plan for IP address allocation and management for your Kubernetes clusters, considering factors like IP address space, public and private IP addresses, and static or dynamic allocation.

Network security

Implement network security measures, such as network segmentation, firewalls, and network policies, to protect your multi-cloud Kubernetes environment from unauthorized access and potential threats.

You can also integrate with cloud provider networking services to leverage the networking services offered by your chosen cloud providers, such as **Virtual Private Cloud (VPC)**, VPC peering, Private Link, or Direct Connect, to establish secure and efficient connectivity between your Kubernetes clusters and other cloud resources.

Network monitoring and visibility

Implement monitoring solutions to gain insights into your multi-cloud Kubernetes network, including traffic patterns, performance metrics, and potential bottlenecks or security threats.

Network resilience and high availability

Design your network architecture with resilience and high availability in mind, ensuring that your multi-cloud Kubernetes environment can withstand network failures or outages without impacting application performance.

By addressing these networking requirements, you can create a robust, secure, and efficient multi-cloud Kubernetes environment that supports seamless communication between applications and services across different cloud providers.

Planning for security and compliance

Planning for security and compliance in a multi-cloud Kubernetes environment is essential to protect your applications, data, and infrastructure. Here are some key aspects to consider when addressing security and compliance:

- **Authentication and authorization**: Implement strong authentication and authorization mechanisms to ensure that only authorized users can access your Kubernetes clusters and resources. This includes using **role-based access control (RBAC)**, integrating with identity providers like LDAP or Active Directory, and leveraging cloud provider-specific **Identity and Access Management (IAM)** services.
- **Network security**: Configure network security measures such as firewalls, network segmentation, and network policies to restrict traffic between resources and prevent unauthorized access to your Kubernetes environment. Additionally, use encryption for data in transit, such as TLS for secure communication between components.
- **Container security**: Ensure the security of your container images and runtime environment. This includes using trusted base images, scanning images for vulnerabilities, implementing image signing and verification, and following best practices for container runtime security.
- **Data security**: Protect your data at rest using encryption and access controls. Leverage encryption features provided by your cloud providers, such as server-side encryption for storage services, and manage encryption keys securely using key management services.
- **Compliance with industry-specific regulations**: Identify the compliance requirements for your industry, such as GDPR, HIPAA, PCI DSS, or other data protection and privacy regulations. implement the necessary controls and processes to ensure your multi-cloud Kubernetes environment adheres to these requirements.
- **Security monitoring and auditing**: Establish monitoring and auditing solutions to detect and respond to security incidents in your multi-cloud Kubernetes environment. This includes monitoring logs, setting up intrusion detection systems, and conducting regular security audits.
- **Security policies and governance**: Develop and enforce security policies and governance practices within your organization to maintain the security and compliance of your multi-cloud Kubernetes environment. This may involve creating guidelines for managing secrets, configuring security groups, and patching vulnerabilities.
- **Incident response and recovery**: Plan for incident response and recovery, including defining roles and responsibilities, establishing communication channels, and creating playbooks for responding to security incidents.
- **Regular security assessments**: Conduct regular security assessments to identify vulnerabilities and risks in your multi-cloud Kubernetes environment. This can include vulnerability scanning, penetration testing, and infrastructure reviews.
- **Employee training and awareness**: Invest in security training and awareness programs for your team to ensure they understand the importance of security and

compliance in the multi-cloud Kubernetes environment and follow best practices to mitigate risks.

By addressing these security and compliance aspects, you can create a robust and secure multi-cloud Kubernetes environment that meets industry regulations and protects your applications, data, and infrastructure from potential threats.

Designing the Kubernetes cluster

Designing a Kubernetes cluster for a multi-cloud environment requires careful planning and consideration of various aspects, such as cluster size, node configuration, networking, and storage. Here are the key steps to design a Kubernetes cluster for a multi-cloud setup:

Define cluster objectives

Start by identifying the objectives you aim to achieve with your Kubernetes cluster, such as high availability, scalability, cost optimization, or specific performance requirements.

- **Cluster size and topology**: Determine the number of clusters you need across your cloud providers, based on your objectives, workload distribution, and fault tolerance requirements. You may choose to have a single cluster per cloud provider or multiple clusters per provider. Requirements may also dictate many clusters on one provider with single clusters in others. Your specific requirements will guide you in this decision.
- **Node configuration**: Design the configuration of your nodes, including the types of instances or virtual machines to use, the amount of CPU and memory, and the operating system. Consider factors like performance requirements, cost optimization, and compatibility with your workloads and cloud provider services.
- **Control plane configuration**: In designing a multi-cloud Kubernetes environment, the control plane configuration involves determining the architecture and setup of key Kubernetes control plane components. This includes decisions on whether to deploy a single or multi-master configuration for high availability, and considerations for distributing control plane components across multiple clouds or regions. The aim is to establish a resilient, well-managed control plane that orchestrates and maintains the cluster's overall health, ensuring consistent and reliable operations across the multi-cloud environment.
- **Networking**: Select an appropriate **Container Network Interface (CNI)** plugin that supports your requirements and is compatible with your cloud providers. Plan for intra-cluster networking, inter-cluster networking, load balancing, DNS configuration, IP address management, and network security.
- **Storage**: Design your storage solution, considering factors like data persistence, performance, and cost. Leverage the storage services provided by your cloud

providers, such as block storage, file storage, or object storage, and choose the appropriate storage classes and provisioners for your Kubernetes environment.

- **High availability and disaster recovery**: Implement high availability and disaster recovery strategies for your Kubernetes cluster, such as using multiple availability zones or regions, deploying multi-master control planes, and setting up backup and restore mechanisms.
- **Autoscaling**: In the design of a multi-cloud Kubernetes environment, cluster autoscaling is a crucial feature aimed at optimizing resource utilization and adapting to varying workload demands. This involves dynamically adjusting the number of worker nodes in the cluster based on real-time resource requirements. By automatically scaling the cluster up during periods of increased demand and down during lower utilization, cluster autoscaling ensures efficient use of cloud resources, minimizes costs, and enhances the environment's ability to handle fluctuating workloads across multiple cloud providers, contributing to a more responsive and cost-effective infrastructure.
- **Monitoring and logging**: Design a monitoring and logging solution that provides visibility into the health and performance of your Kubernetes cluster, including metrics like CPU and memory usage, network traffic, and container logs.
- **Security and compliance**: Ensure that your Kubernetes cluster design addresses security and compliance requirements, such as authentication and authorization, network security, container security, data encryption, and adherence to industry-specific regulations.

By carefully designing your Kubernetes cluster with these aspects in mind, you can create a robust, scalable, and efficient multi-cloud environment that meets your organization's objectives and requirements.

Establish monitoring and logging

Establishing monitoring and logging in a multi-cloud Kubernetes environment is essential for maintaining visibility into the performance, health, and security of your applications and infrastructure. Here are the key steps to set up monitoring and logging for your multi-cloud Kubernetes setup:

- **Centralized logging**: Implement a centralized logging solution to aggregate logs from your Kubernetes clusters across different cloud providers. This may involve using managed logging services like Google Cloud Operations (formerly Stackdriver), Amazon CloudWatch Logs, or Azure Monitor Logs. Alternatively, you can set up open-source solutions such as **Elasticsearch, Logstash, and Kibana (ELK)** stack or Grafana Loki for log aggregation and visualization.
- **Application logging**: Ensure that your applications are configured to generate logs in a standardized format, such as JSON, and include relevant information like

timestamps, log levels, and contextual metadata. This makes it easier to analyze and troubleshoot application issues.

- **Container-level logging**: Configure your Kubernetes clusters to collect logs from containers using log drivers or sidecar containers. These logs can include stdout and stderr streams, as well as application-specific log files.

- **Kubernetes system component logs**: Collect logs from Kubernetes system components, such as the API server, etcd, kubelet, and controller manager, to gain insights into the health and performance of your Kubernetes control plane.

- **Monitoring metrics**: In the design of a multi-cloud Kubernetes environment, monitoring metrics plays a pivotal role in maintaining visibility and ensuring the health and performance of the infrastructure. This involves setting up a comprehensive monitoring system that captures key metrics such as CPU usage, memory utilization, network traffic, and application-specific performance indicators. By leveraging tools like Prometheus, along with data visualization tools like Grafana, or cloud provider-native monitoring services, organizations can gain insights into the behavior of their multi-cloud Kubernetes environment, enabling proactive issue detection, efficient resource management, and informed decision-making for continuous optimization and enhancement of the infrastructure's overall performance.

- **Kubernetes-native monitoring**: Leverage Kubernetes-native monitoring tools like Metrics Server, which provides resource usage metrics for nodes and pods, and kube-state-metrics, which exposes Kubernetes object state metrics.

- **Network monitoring**: Monitor network traffic and performance in your multi-cloud Kubernetes environment, including intra-cluster and inter-cluster communication. This may involve using network monitoring tools like Weave Scope, Calico, or Cilium.

- **Alerting and notifications**: Configure alerting and notification mechanisms to proactively inform you of potential issues or performance anomalies in your Kubernetes environment. This may involve integrating your monitoring and logging solutions with notification services like *PagerDuty, Slack*, or email.

- **Dashboards and visualization**: Create dashboards and visualizations to display **key performance indicators (KPIs)**, metrics, and logs for your multi-cloud Kubernetes environment. This can help you quickly identify trends, spot issues, and gain insights into your applications and infrastructure.

- **Continuous monitoring and improvement**: Regularly review and update your monitoring and logging configuration to ensure that you are capturing the necessary data to maintain visibility into your multi-cloud Kubernetes environment. Continuously analyze the collected data to identify areas for optimization and improvement.

By establishing monitoring and logging for your multi-cloud Kubernetes environment, you can maintain visibility into your applications and infrastructure, quickly identify and troubleshoot issues, and ensure the performance, health, and security of your containerized applications.

Defining backup and disaster recovery strategies

Defining backup and disaster recovery strategies for a multi-cloud Kubernetes environment involves planning for data protection, system redundancy, and rapid recovery in the event of failures or disasters. Here are the key steps to establish effective backup and disaster recovery strategies:

1. **Identify critical components**: Determine the critical components of your multi-cloud Kubernetes environment, such as control plane components, worker nodes, application data, and configuration data, that require backup and protection.
2. **Data backup**: Implement a data backup strategy to protect your application data, including databases, object storage, and other persistent data. Leverage the backup solutions provided by your cloud providers or use third-party backup tools like Velero, Kasten K10, or Restic.
3. **Configuration backup**: Back up your Kubernetes cluster configuration, such as resource manifests, Helm charts, and **custom resource definitions (CRDs)**, to ensure that you can recreate your environment in the event of a disaster. Store these backups in a secure and version-controlled repository.
4. **Control plane backup**: Ensure that your Kubernetes control plane components, such as etcd data, API server configuration, and other control plane components, are backed up regularly. This may involve using etcd snapshotting, cloud provider-managed Kubernetes services with built-in backup capabilities, or third-party backup tools.
5. **High availability and redundancy**: Design your multi-cloud Kubernetes environment with high availability and redundancy in mind. This includes deploying your control plane components and worker nodes across multiple availability zones or regions and using multi-master control plane configurations.
6. **Load balancing and failover**: Implement load balancing and failover mechanisms to distribute traffic across your services and ensure high availability. This may involve using Kubernetes Ingress controllers, cloud provider-managed load balancers, or third-party load balancing solutions.
7. **Disaster recovery plan**: Develop a comprehensive disaster recovery plan that includes the steps to recover your Kubernetes environment, applications, and data in the event of a failure or disaster. This plan should cover the recovery of control plane components, worker nodes, application data, and configuration data.
8. **Test your disaster recovery procedures**: Regularly test your disaster recovery procedures to ensure that they are effective and that your team is familiar with the steps to recover your environment in the event of a disaster. This may involve conducting recovery drills or simulating failure scenarios.

9. **Monitor and update your strategies**: Continuously monitor and update your backup and disaster recovery strategies to ensure that they remain effective as your multi-cloud Kubernetes environment evolves. This may involve adjusting backup schedules, modifying recovery procedures, or incorporating new cloud provider services and features.

By defining and implementing backup and disaster recovery strategies for your multi-cloud Kubernetes environment, you can minimize the impact of failures or disasters on your applications and ensure that your environment can be quickly and effectively recovered.

Set up CI/CD pipelines

Setting up CI/CD pipelines for a multi-cloud Kubernetes environment helps automate the process of building, testing, and deploying applications. Here are the key steps to set up CI/CD pipelines:

1. **Choose CI/CD tools**: Select appropriate CI/CD tools that meet your requirements and integrate well with your multi-cloud Kubernetes environment. Common choices include Jenkins, GitLab CI/CD, GitHub Actions, CircleCI, Travis CI, or cloud provider-specific services like Google Cloud Build, AWS CodePipeline, and Azure Pipelines. There are also options for Kubernetes specific tooling such as ArgoCD or Flux.

2. **Source code management**: Use a version control system like Git to manage your application source code, configuration files, and resource manifests. Integrate your CI/CD tools with your version control system to automatically trigger pipelines when changes are pushed or merged.

3. **Containerization**: Containerize your applications using Docker or another container runtime. Write Dockerfiles to define your application images and dependencies, ensuring that they are built consistently across all environments.

4. **Build and test**: Configure your CI/CD pipeline to build your application and run unit tests, integration tests, and other automated tests as needed. Ensure that your pipeline fails fast in case of test failures, preventing the deployment of broken code.

5. **Container registry**: Push your container images to a container registry that is accessible from your multi-cloud Kubernetes environment. This may involve using cloud provider-managed registries like Google Container Registry, Amazon Elastic Container Registry, Azure Container Registry, or third-party registries like Docker Hub or Quay.io.

6. **Deploy to Kubernetes**: Automate the deployment of your applications to your multi-cloud Kubernetes environment using tools like Helm, ArgoCD or Flux.

Configure your CI/CD pipeline to deploy your applications using Kubernetes manifests directly or Helm charts. Each solution will have its own deployment patterns so use your requirements as a guide to make this choice.

7. **Rollout strategies**: Implement appropriate rollout strategies for your application deployments, such as rolling updates, blue-green deployments, or canary releases. Use Kubernetes-native features like Deployments, StatefulSets, and Jobs, or third-party tools like Flagger or Argo Rollouts.
8. **Monitoring and feedback**: Monitor the performance and health of your applications in your multi-cloud Kubernetes environment and provide feedback to your development team. Integrate your CI/CD pipeline with monitoring and alerting tools like Prometheus, Grafana, and Alertmanager to ensure that issues are detected and addressed quickly.
9. **Continuous improvement**: Continuously iterate and improve your CI/CD pipeline, incorporating new tools, practices, and feedback to optimize your application development and deployment process.

By setting up CI/CD pipelines for your multi-cloud Kubernetes environment, you can automate the process of building, testing, and deploying applications, ensuring that your applications are consistently delivered and updated across all environments.

Investing in training and skill development

Investing in training and skill development is essential for the success of your multi-cloud Kubernetes environment. By empowering your team with the necessary skills and knowledge, you can improve efficiency, reduce errors, and drive innovation. Here are some strategies to invest in training and skill development:

- **Assess current skills:** Start by assessing the current skill set of your team members. Identify gaps and areas for improvement related to Kubernetes, cloud provider services, containerization, CI/CD pipelines, and other relevant topics.
- **Create a training plan**: Develop a comprehensive training plan that addresses the identified skill gaps and aligns with your organization's objectives and goals. This plan should include a mix of formal training, self-paced learning, and practical hands-on experience.
- **Formal training courses**: Enroll your team in formal training courses offered by cloud providers, Kubernetes training partners, or other specialized training providers. These courses typically cover a wide range of topics, from beginner to advanced levels, and can lead to industry-recognized certifications.
- **Online resources**: Encourage your team to leverage online resources for self-paced learning, such as blogs, tutorials, webinars, and videos. Platforms like Coursera, Pluralsight, Udemy, edX, and the official Kubernetes website offer a wealth of learning materials.

- **Hands-on experience**: Provide opportunities for your team to gain hands-on experience with multi-cloud Kubernetes environments. This can include setting up sandbox environments, participating in hackathons or workshops, or contributing to open-source projects.
- **Knowledge sharing sessions**: Organize regular knowledge sharing sessions within your team, where members can present their learnings, share best practices, and discuss challenges. This can foster collaboration and help team members learn from one another.
- **Mentoring and coaching**: Encourage experienced team members to mentor and coach less experienced colleagues. This can help transfer knowledge and skills more effectively and create a supportive learning culture within your team.
- **Conferences and events**: Encourage your team to attend industry conferences and events related to Kubernetes, cloud computing, and other relevant topics. These events provide opportunities to learn from experts, network with peers, and stay up-to-date with the latest trends and technologies.
- **Continuous learning**: Foster a culture of continuous learning within your organization by setting clear expectations for ongoing skill development, providing time and resources for learning, and recognizing and rewarding learning achievements.
- **Evaluate and adjust**: Regularly evaluate the effectiveness of your training and skill development initiatives. Gather feedback from your team members, assess their progress, and adjust your training plan as needed to ensure continuous improvement.

By investing in training and skill development for your team, you can build a strong foundation of knowledge and skills that will enable your organization to effectively manage and optimize your multi-cloud Kubernetes environment.

Setting up your local development environment

Setting up a local development environment for your DevOps team members involves providing them with the necessary tools, configurations, and access to collaborate effectively on your project. Here is a general approach to setting up a local development environment for DevOps team members:

- **Standardize the environment**: Ensure that all team members are using the same OS and version or a compatible one to avoid compatibility issues when developing and deploying applications. If needed, consider using **virtual machine**s (**VMs**) or containers to create a consistent environment across different machines.
- **Version control**: Set up a version control system like Git and provide team members with access to your project repositories. Ensure they follow best practices for branching, committing, and merging code.

- **Integrated Development Environment (IDE)**: Choose a suitable IDE or text editor for your team members based on the programming languages and frameworks used in the project. Examples include Visual Studio Code, IntelliJ IDEA, or PyCharm. Ensure the IDE is configured with necessary plugins and extensions to support your development stack.
- **Language runtime and SDKs**: Install the appropriate language runtime and SDKs for your project, such as Python, Java, or Node.js, along with the required versions.
- **Dependency management**: Use dependency management tools like *npm*, *pip*, or *Maven* to manage project dependencies and ensure consistency across team members' local environments.
- **Local development tools**: Install and configure tools that streamline local development and testing, such as Docker, Minikube, or Skaffold. This allows team members to work with containerized applications and Kubernetes clusters locally, ensuring a consistent experience between local and production environments.
- **Code linting and formatting**: Set up code linting and formatting tools, like ESLint, Pylint, or Black, to enforce coding standards and best practices across the team.
- **Continuous Integration (CI) tools**: Integrate your version control system with CI tools like Jenkins, GitLab CI, or GitHub Actions, so that team members can automatically test and validate their changes before merging them into the main branch.
- **Documentation**: Provide team members with access to project documentation, including architecture diagrams, API documentation, and coding guidelines, to ensure they have a clear understanding of the project's requirements and design.
- **Communication and collaboration tools**: Set up communication and collaboration tools like Slack, Microsoft Teams, or Jira to enable team members to communicate effectively, track progress, and share knowledge.

By providing a well-configured and standardized local development environment for your DevOps team members, you can enhance productivity, improve code quality, and facilitate collaboration across the team.

Conclusion

In conclusion, comprehensive planning and preparation are crucial steps before embarking on the journey to build a multi-cloud Kubernetes environment. In this chapter we addressed key aspects such as business and technical objectives, infrastructure components, management and orchestration tools, security and compliance requirements, and cost management strategies, you can lay a solid foundation for a successful and efficient deployment. Investing time and effort in the planning phase helps mitigate risks, optimize resource allocation, and ensure alignment with your organization's goals. Moreover, it fosters better collaboration among team members and stakeholders, leading to a more maintainable and resilient multi-cloud Kubernetes infrastructure that effectively leverages the unique advantages offered by different cloud providers. Ultimately, diligent planning and preparation are essential in unlocking the full potential of a multi-cloud Kubernetes environment, enabling your organization to thrive in today's dynamic and competitive landscape.

Get ready to dive into the exciting world of running Kubernetes on multiple cloud providers in our upcoming chapter! As you navigate through the complexities of deploying and managing your Kubernetes clusters across diverse cloud platforms, we will guide you step by step, exploring best practices, powerful tools, and proven strategies to maximize the benefits of a multi-cloud environment. You will learn how to seamlessly integrate your Kubernetes infrastructure, streamline your application deployments, and achieve unparalleled scalability, resilience, and cost optimization.

Join our book's Discord space

Join the book's Discord Workspace for Latest updates, Offers, Tech happenings around the world, New Release and Sessions with the Authors:

https://discord.bpbonline.com

CHAPTER 2
Running Multi-Cloud Kubernetes

Introduction

This chapter will dive deep into the benefits and challenges of running multiple Kubernetes clusters across various public cloud environments. We will explore the reasons behind adopting a multi-cloud strategy, the evolution of multi-cloud Kubernetes, and the key considerations when implementing a multi-cloud Kubernetes environment. By the end of this chapter, you will have a comprehensive understanding of the advantages, potential pitfalls, and critical factors to keep in mind when managing Kubernetes clusters in multiple public clouds.

Structure

The chapter covers the following topics:
- Rationale behind multi-cloud Kubernetes
- Evolution of multi-cloud Kubernetes
- Pros of multi-cloud Kubernetes environment
- Cons of multi-cloud Kubernetes environment
- Multi-cloud Kubernetes considerations
- Hypothetical use cases

- Hypothetical case study
- Industry best practices

Objectives

This chapter will equip you with a thorough understanding of the pros and cons of running multiple Kubernetes clusters in multiple public clouds, enabling you to make well-informed decisions and design a robust, efficient, and secure multi-cloud Kubernetes environment tailored to your organization's unique requirements.

Rationale behind multi-cloud Kubernetes

Having a multi-cloud Kubernetes strategy can provide numerous benefits and help solve various problems that organizations face when operating in a single-cloud environment. By deploying Kubernetes clusters across multiple cloud providers, organizations can address several challenges and achieve greater flexibility, resilience, and efficiency. Here are some key problems that a multi-cloud Kubernetes approach can solve:

- **Vendor lock-in avoidance**: Relying on a single cloud provider can lead to vendor lock-in, making it difficult to switch providers or take advantage of better offerings from competitors. A multi-cloud Kubernetes strategy helps to minimize this risk by diversifying cloud provider dependencies.
- **Increased redundancy and high availability**: Deploying Kubernetes clusters across multiple cloud providers ensures greater redundancy and higher availability for applications. In the event of a cloud provider outage or service disruption, the multi-cloud environment can continue to operate with minimal impact on the overall system.
- **Improved geographical coverage and latency**: A multi-cloud Kubernetes approach allows organizations to deploy clusters closer to their end-users, reducing latency and improving user experience. This is especially useful for organizations with a global user base or specific regional requirements.
- **Leveraging unique features and capabilities:** Different cloud providers offer unique features, services, and pricing options. A multi-cloud strategy allows organizations to benefit from the best of each provider, optimizing their infrastructure and application deployments based on specific needs.
- **Cost optimization and negotiation power:** By using multiple cloud providers, organizations can optimize their costs by taking advantage of the most cost-effective services and pricing models. This approach also provides increased negotiation power when discussing contracts and discounts with cloud providers.
- **Enhanced security and compliance**: A multi-cloud Kubernetes environment allows organizations to distribute their workloads and data across multiple cloud

providers, reducing the potential impact of security breaches and ensuring better compliance with regional and industry-specific regulations.

- **Flexibility in scaling and resource management**: Deploying Kubernetes clusters in multiple clouds provides greater flexibility in scaling applications and managing resources based on demand, cost, and performance requirements. This can lead to more efficient resource utilization and better overall performance.
- **Supporting multi-tenant and multi-region deployments**: A multi-cloud strategy enables organizations to better support multi-tenant and multi-region deployments, providing improved isolation between tenants and ensuring compliance with data sovereignty requirements.
- **Disaster recovery and business continuity**: With Kubernetes clusters deployed across multiple cloud providers, organizations can implement more robust disaster recovery and business continuity plans, ensuring minimal downtime and data loss in the event of a disaster or cloud provider outage.

By adopting a multi-cloud Kubernetes approach, organizations can address these problems and gain a competitive edge by leveraging the strengths of different cloud providers, improving the resilience and performance of their applications, and optimizing costs. Refer to the following figure for a single cloud Kubernetes architecture:

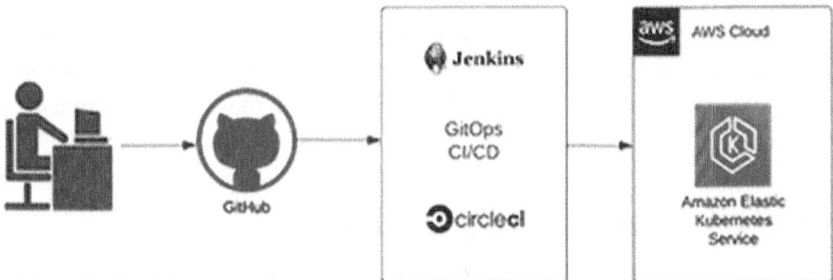

Figure 2.1: A single cloud Kubernetes architecture

Evolution of multi-cloud Kubernetes

Let us take a quick step back in time to help set a foundation before we move forward, this is the journey that got us to where we are now.

To begin, Kubernetes, also known as *K8s*, is an open-source container orchestration platform that automates the deployment, scaling, and management of containerized applications. Containers help developers package applications and their dependencies together, making it easier to deploy and maintain applications consistently across different environments. Kubernetes manages these containers, ensuring that applications run smoothly and efficiently. A multi-cloud environment, on the other hand, refers to the use of multiple cloud computing services from different providers, such as AWS, Google Cloud, Azure, and others. Organizations may choose to adopt a multi-cloud strategy for various reasons, including avoiding vendor lock-in, achieving greater redundancy, improving geographical coverage, and leveraging unique features and capabilities of different cloud providers. Now that we have a basic understanding of Kubernetes and multi-cloud environments, let us discuss the evolution of multi-cloud Kubernetes. Refer to the following figure for multi-cloud Kubernetes architecture:

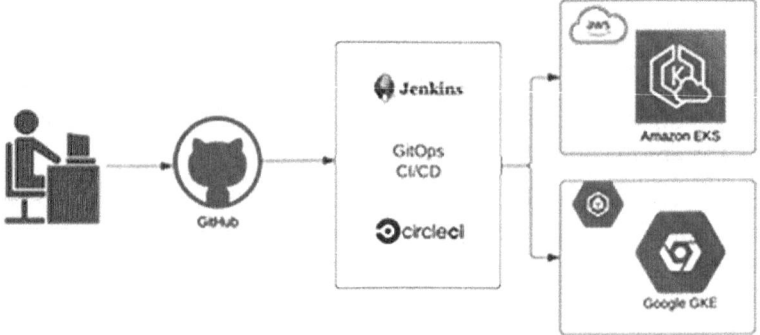

Figure 2.2: Multi-cloud Kubernetes architecture

Pre-Kubernetes era

Before Kubernetes, organizations had limited options for managing containers in multi-cloud environments. They had to rely on manual processes or use multiple tools to deploy and manage containers across different cloud platforms, leading to increased complexity and inefficiencies.

Kubernetes emergence (2014)

Kubernetes was initially developed by Google as an internal project called **Borg**, which managed Google's large-scale container workloads. In 2014, Google open-sourced Kubernetes, making it available to the broader community. This marked the beginning of a new era in container orchestration, as Kubernetes quickly gained popularity due to its powerful features, scalability, and extensibility.

Managed Kubernetes services (2017 onward)

As Kubernetes became increasingly popular, major cloud providers started offering managed Kubernetes services. These services simplify the process of deploying and managing Kubernetes clusters by handling many administrative tasks, such as cluster provisioning, upgrades, and scaling. Some of the first managed Kubernetes services were **Google Kubernetes Engine (GKE)**, **Amazon Elastic Kubernetes Service (EKS)**, and **Azure Kubernetes Service (AKS)**.

Multi-cloud Kubernetes tools and solutions

With the rise of managed Kubernetes services, various tools and solutions were developed to simplify the management of Kubernetes clusters in multi-cloud environments. Some popular tools include **Kubernetes Operations** (**kops**), Kustomize, ArgoCD and Flux. These tools enable organizations to deploy, manage, and synchronize Kubernetes clusters across multiple cloud providers more efficiently.

Multi-cloud service mesh and networking solutions

As organizations began adopting multi-cloud Kubernetes environments, the need for better networking solutions and service meshes arose. Projects like Istio, Linkerd, and Consul addressed these challenges by providing advanced traffic management, security, and observability features for services running in Kubernetes clusters across multiple clouds.

GitOps and Infrastructure as Code

To further streamline the management of multi-cloud Kubernetes environments, organizations began adopting GitOps and **Infrastructure as Code (IaC)** practices. These practices enable organizations to manage their infrastructure using code, stored in version control systems like Git, and automate the deployment and management of Kubernetes clusters across multiple cloud providers. Tools like Terraform, Pulumi, and Argo CD are widely used in this context.

In summary, the evolution of multi-cloud Kubernetes can be seen as a series of developments that have made it increasingly easier and more efficient for organizations to deploy and manage containerized applications across multiple cloud providers. From the emergence of Kubernetes to the introduction of managed Kubernetes services, and the development of tools and best practices for multi-cloud Kubernetes management; the landscape has continuously evolved, enabling organizations to reap the benefits of multi-cloud Kubernetes environments.

Pros of multi-cloud Kubernetes

Running a multi-cloud Kubernetes environment offers several advantages that can help organizations achieve greater flexibility, resilience, and efficiency in their infrastructure and applications. Here are some key pros of adopting a multi-cloud Kubernetes approach:

- **A common platform across all platforms**: One of the primary advantages of using a single container orchestrator like Kubernetes across multiple cloud providers is the consistency and standardization it offers. Having a common platform simplifies infrastructure management, streamlines processes, and reduces overall complexity. With Kubernetes as the orchestrator, teams can benefit from unified management, easier workload migration, skillset consolidation, streamlined **Continuous Integration and Continuous Deployment (CI/CD)**, and vendor-agnostic infrastructure. This approach enables organizations to focus on developing expertise in Kubernetes, which is transferable across various cloud providers, and promotes agility, flexibility, and operational efficiency.

- **Vendor lock-in avoidance**: By deploying Kubernetes clusters across different cloud providers, organizations can minimize the risk of vendor lock-in. This diversification allows organizations to remain agile, switch providers more easily, and take advantage of better offerings from competitors.

- **Increased redundancy and high availability**: A multi-cloud Kubernetes environment ensures greater redundancy and higher availability for applications. If a cloud provider experiences an outage or service disruption, the multi-cloud environment can continue to operate with minimal impact, reducing the chances of a single point of failure affecting the entire system.

- **Improved geographical coverage and reduced latency**: Deploying Kubernetes clusters across multiple cloud providers allows organizations to have better geographical coverage and reduce latency for their end-users. This is particularly beneficial for organizations with a global user base or specific regional requirements, as they can deploy clusters closer to their users, improving overall user experience.

- **Leveraging unique features and capabilities**: Different cloud providers offer unique features, services, and pricing options. A multi-cloud strategy enables organizations to benefit from the best of each provider, optimizing their infrastructure and application deployments based on specific needs.

- **Cost optimization and negotiation power**: Using multiple cloud providers allow organizations to optimize their costs by taking advantage of the most cost-effective services and pricing models. A multi-cloud approach also provides increased negotiation power when discussing contracts and discounts with cloud providers.

- **Enhanced security and compliance**: A multi-cloud Kubernetes environment enables organizations to distribute their workloads and data across multiple cloud providers, reducing the potential impact of security breaches and ensuring better compliance with regional and industry-specific regulations.

- **Flexibility in scaling and resource management**: Deploying Kubernetes clusters in multiple clouds provides greater flexibility in scaling applications and managing resources based on demand, cost, and performance requirements. This can lead to more efficient resource utilization and better overall performance.

- **Supporting multi-tenant and multi-region deployments**: A multi-cloud strategy enables organizations to better support multi-tenant and multi-region deployments, providing improved isolation between tenants and ensuring compliance with data sovereignty requirements.

- **Disaster recovery and business continuity**: With Kubernetes clusters deployed across multiple cloud providers, organizations can implement more robust disaster recovery and business continuity plans, ensuring minimal downtime and data loss in the event of a disaster or cloud provider outage.

Overall, a multi-cloud Kubernetes approach allows organizations to address various challenges and gain a competitive edge by leveraging the strengths of different cloud providers, improving the resilience and performance of their applications, and optimizing costs.

Cons of multi-cloud Kubernetes

While running a multi-cloud Kubernetes environment offers several advantages, running a multi-cloud Kubernetes environment has several cons and challenges that organizations should be aware of before adopting this approach. Some of the key disadvantages and challenges include:

- **Increased complexity**: Managing Kubernetes clusters across multiple cloud providers adds complexity to the infrastructure. Each cloud provider has its unique set of services, APIs, and tools, which can require additional effort and expertise to manage effectively.
- **Operational overhead**: Multi-cloud Kubernetes environments often involve higher operational overhead. Managing and monitoring clusters across different cloud providers includes handling authentication, security policies, and logging across multiple platforms, which can be time-consuming and resource intensive.
- **Inconsistent features and services**: Different cloud providers offer varying features and services, which may not be entirely compatible or consistent across providers. This inconsistency can create challenges when using specific features or services that are only available on certain platforms or when maintaining a consistent infrastructure across multiple clouds.
- **Integration challenges**: Integrating various tools, services, and APIs from multiple cloud providers can be challenging and may require additional effort to ensure seamless interoperability. This may involve building custom integrations or relying on third-party tools to bridge the gaps between different platforms.
- **Cost management**: While a multi-cloud approach can potentially lead to cost optimization, it can also make cost management more complicated. Tracking and optimizing costs across multiple cloud providers can be difficult, especially when accounting for different pricing models, discounts, and billing cycles.
- **Skills and expertise**: Managing a multi-cloud Kubernetes environment requires a broad range of skills and expertise in different cloud platforms. Finding and retaining talent with the necessary knowledge and experience can be challenging and may increase your organization's overall operational costs.

- **Data transfer costs**: Moving data between different cloud providers can incur additional costs due to data transfer fees. Depending on your data usage patterns and the frequency of data transfers, this can have a significant impact on your overall expenses.
- **Latency and performance**: Deploying Kubernetes clusters across multiple cloud providers can improve geographical coverage and reduce latency. However, it can also introduce performance challenges. Network latency between different cloud providers can affect the performance of applications and services, especially if they rely on communication between clusters in different clouds.
- **Compliance and regulatory challenges**: Managing compliance and regulatory requirements across multiple cloud providers can be complex. Ensuring data sovereignty, privacy, and industry-specific regulations are met consistently across all platforms may require additional effort and resources.

Organizations need to carefully consider these potential challenges and weigh the pros and cons before deciding on a multi-cloud Kubernetes strategy. With proper planning, management, and tooling, many of these downsides can be mitigated, allowing organizations to successfully deploy a multi-cloud Kubernetes environment.

Multi-cloud Kubernetes considerations

When planning and implementing a multi-cloud Kubernetes environment, there are several key considerations to keep in mind to ensure a successful deployment. These include:

- **Strategy and goals**: Define the overall strategy and goals for your multi-cloud Kubernetes environment. Determine the specific reasons for adopting a multi-cloud approach, such as avoiding vendor lock-in, increasing redundancy, improving geographical coverage, or leveraging unique features and capabilities of different cloud providers.
- **Infrastructure consistency**: Strive to maintain consistency in your infrastructure across multiple cloud providers. This may involve using common Kubernetes configurations, container images, and deployment processes to reduce complexity and minimize potential issues.
- **Networking and connectivity**: Plan the networking and connectivity between your Kubernetes clusters in different cloud providers. Consider using tools and technologies such as **Virtual Private Networks (VPNs)**, direct connections, or multi-cloud networking solutions to establish secure, high-performance connections between your clusters. //insert diagram.
- **Security and compliance**: Ensure that your multi-cloud Kubernetes environment meets your organization's security and compliance requirements. This includes managing access controls, encryption, network security, and adhering to data sovereignty and industry-specific regulations across all cloud providers.

- **Monitoring and observability**: Implement comprehensive monitoring and observability solutions that cover your entire multi-cloud Kubernetes environment. This includes gathering metrics, logs, and traces from clusters across all cloud providers and consolidating them into a unified monitoring system for better visibility and analysis.
- **Cost management**: Establish cost management strategies and processes to track and optimize expenses across multiple cloud providers. This may involve using cost management tools, setting budget limits, and regularly reviewing and adjusting resource allocations based on usage patterns.
- **Automation and tooling**: Leverage automation and tooling to streamline the deployment and management of your multi-cloud Kubernetes environment. Tools like Terraform, kops, Cluster API, and service meshes can help simplify the process of managing Kubernetes clusters across multiple cloud providers.
- **Disaster recovery and business continuity**: Develop a robust disaster recovery and business continuity plan that takes into account the multi-cloud nature of your Kubernetes environment. This includes implementing backup and recovery strategies, as well as planning for failover and redundancy across different cloud providers.
- **Skills and training**: Ensure that your team has the necessary skills and expertise to manage a multi-cloud Kubernetes environment. This may involve investing in training and professional development, as well as hiring or partnering with experts who have experience in managing Kubernetes deployments across different cloud providers.

By taking these key considerations into account, you can build a multi-cloud Kubernetes environment that meets your organization's goals and requirements while minimizing potential challenges and complexities.

Hypothetical use cases

Here are five of the most common and high priority use cases for multi-cloud Kubernetes. They include requirements, potential challenges and desired outcomes as well which will prove useful in any justification for multi-cloud Kubernetes environment. The use cases are as follows:

Use case 1: High availability and redundancy

Requirements: The company requires a highly available infrastructure to minimize downtime and ensure continuous service to its customers.

The company wants to mitigate risks associated with relying on a single cloud provider, such as service outages or regional failures.

Potential challenges: Implementing a multi-cloud Kubernetes environment for high availability and redundancy introduces challenges such as ensuring interoperability and portability across diverse cloud providers, managing complex networking configurations, addressing data consistency and synchronization issues, coordinating identity and access management, optimizing costs while avoiding over-provisioning, adopting unified tooling and automation, aligning service level agreements, handling operational complexities, mitigating vendor lock-in risks, and developing robust disaster recovery plans. Overcoming these challenges requires meticulous planning, a deep understanding of multi-cloud dynamics, and the adoption of best practices to maintain a reliable, resilient infrastructure that meets the company's uptime and service continuity goals.

Desired outcomes: By deploying Kubernetes clusters across multiple cloud providers, the company achieves higher availability and redundancy.

The risk of service outages or regional failures is reduced, ensuring better reliability and continuous service for customers.

Use case 2: Improved geographical coverage and latency reduction

Requirements: The company has a global customer base and needs to provide low-latency access to its services.

The company wants to optimize performance for users in different regions by deploying infrastructure closer to them.

Potential challenges: Implementing Kubernetes clusters across multiple cloud providers for improved geographical coverage and latency reduction presents challenges related to orchestrating seamless communication and data synchronization across dispersed clusters. Coordinating the deployment of clusters in diverse regions while ensuring consistent performance requires addressing networking complexities, latency variations, and data consistency challenges. Efficiently managing and optimizing traffic routing to direct users to the nearest cluster, as well as maintaining synchronization of data in a globally distributed environment, are critical aspects. Moreover, handling potential differences in cloud provider capabilities, availability, and compliance regulations across regions adds complexity to achieving a uniformly low-latency experience for the global customer base. Successful navigation of these challenges is essential for realizing the desired outcomes of enhanced geographical coverage and improved user experience.

Desired outcomes: Deploying Kubernetes clusters across multiple cloud providers with data centers in various regions allows the company to improve geographical coverage.

Reduced latency leads to better user experience and customer satisfaction.

Use case 3: Cost optimization and flexibility

Requirements: The company wants to optimize its infrastructure costs by leveraging different pricing models, discounts, and features offered by various cloud providers.

The company wants the flexibility to choose the best services and resources for its needs without being locked into a single provider.

Potential challenges: Embracing a multi-cloud Kubernetes environment for cost optimization and flexibility introduces challenges associated with managing diverse pricing models, services, and features across different cloud providers. Variations in billing structures, discount programs, and resource pricing necessitate careful cost analysis and monitoring to ensure effective optimization. Coordinating resource allocation and workload distribution while navigating differences in service offerings and compatibility requires continuous effort to maintain flexibility. Additionally, challenges may arise in achieving a cohesive and unified management approach across diverse cloud environments, necessitating the adoption of consistent tooling and practices. Balancing the pursuit of cost savings with the need for operational simplicity and agility presents an ongoing challenge in realizing the desired outcomes of optimized infrastructure costs and enhanced flexibility.

Desired outcomes: A multi-cloud Kubernetes environment enables the company to take advantage of cost savings and unique features from different cloud providers.

The company gains the flexibility to allocate resources and choose services that best meet its requirements, optimizing costs and performance.

Use case 4: Disaster recovery and business continuity

Requirements: The company needs a robust disaster recovery plan to minimize the impact of potential data loss or service disruptions.

The company wants to ensure business continuity in the event of a disaster or outage affecting one of its cloud providers.

Potential challenges: Implementing a multi-cloud Kubernetes environment for disaster recovery and business continuity introduces challenges related to orchestrating a seamless failover strategy across diverse cloud providers. Ensuring data consistency and synchronization, maintaining application performance, and coordinating the failover process in the event of a disaster or outage require careful planning. Challenges may arise in aligning networking configurations, managing identity and access controls, and mitigating potential differences in service capabilities across clouds. Additionally, orchestrating a robust disaster recovery plan that spans multiple providers necessitates comprehensive testing and validation to guarantee swift and effective failover. The complexity lies in harmonizing these elements to achieve the desired outcomes of a resilient disaster recovery strategy and seamless business continuity in the face of potential disruptions.

Desired outcomes: Deploying Kubernetes clusters across multiple cloud providers enables the company to create a more resilient disaster recovery strategy.

The multi-cloud approach ensures business continuity by allowing the company to quickly failover to another cloud provider in case of an outage.

Use case 5: Compliance and regulatory requirements

Requirements: The company operates in an industry with strict compliance and regulatory requirements, such as data sovereignty or privacy regulations.

The company needs to store and process data in specific regions or according to specific rules.

Potential challenges: Adopting a multi-cloud Kubernetes environment to meet compliance and regulatory requirements presents challenges associated with navigating intricate data sovereignty and privacy regulations. Ensuring that data is stored and processed in accordance with specific regional rules demands meticulous attention to the intricacies of each cloud provider's services, their data governance policies, and the evolving landscape of regulatory requirements. Challenges may emerge in harmonizing identity and access controls, encryption practices, and auditability features across diverse cloud environments. Achieving consistency in compliance measures while balancing the need for operational efficiency and flexibility poses a complex task. The company must stay abreast of evolving regulations, regularly update its policies, and invest in continuous monitoring and auditing to fulfill the desired outcomes of a compliant multi-cloud Kubernetes environment, reducing the risk of penalties or fines associated with regulatory non-compliance.

Desired outcomes: A multi-cloud Kubernetes environment allows the company to select cloud providers that meet its compliance and regulatory requirements.

The company can store and process data in the required regions or according to specific rules, ensuring compliance and reducing the risk of penalties or fines.

Hypothetical case study

Here we will look at a hypothetical case study of Global HealthTech, a company that does not actually exist but has a very real problem they must solve. Let us dive in!

Company: Global HealthTech

Industry: Healthcare technology

Problem statement: Global HealthTech is a leading healthcare technology company that develops and manages a variety of digital health applications and services. Their products are used by healthcare providers, patients, and researchers worldwide. The company's infrastructure is initially deployed on a single cloud provider, which led to concerns about vendor lock-in, potential service outages, and limited geographical coverage. Let us look at how we can approach this case study:

- **Requirements**
    - High availability and redundancy to ensure continuous service for customers.
    - Improved geographical coverage to provide better performance and low-latency access to users worldwide.
    - Compliance with strict data privacy and sovereignty regulations due to the sensitive nature of healthcare data.
    - Scalability to support the growing user base and the increasing volume of data.
- **Objectives**
    - Implement a multi-cloud Kubernetes environment to mitigate risks associated with a single cloud provider.
    - Achieve better redundancy, availability, and geographical coverage for the global user base.
    - Ensure compliance with data privacy and sovereignty regulations across different regions.
    - Create a scalable and flexible infrastructure that can handle the company's growth.
- **Scope**
    - Evaluate and select multiple cloud providers that meet the company's requirements for availability, coverage, and compliance.
    - Design and deploy Kubernetes clusters across the chosen cloud providers.
    - Establish secure networking and connectivity between the clusters.
    - Implement monitoring, logging, and observability solutions for the multi-cloud environment.
    - Develop and maintain a disaster recovery and business continuity plan.
    - Train the team to manage and operate the multi-cloud Kubernetes environment effectively.
    - SSO integration for cluster access
    - Global load balancing will help reduce end user latency and direct requests to the closest geographically located cluster

- **Desired outcomes**
  - By adopting a multi-cloud Kubernetes environment, Global HealthTech successfully mitigates the risks associated with vendor lock-in and single cloud provider outages.
  - The company achieves higher availability and redundancy, ensuring better reliability and continuous service for customers.
  - The multi-cloud approach allows Global HealthTech to adhere to strict data privacy and sovereignty regulations across different regions, reducing the risk of penalties or fines.
  - The infrastructure is more scalable and flexible, enabling the company to accommodate its growing user base and increasing data volume efficiently.
  - Improved geographical coverage and reduced latency lead to better user experience and customer satisfaction.

Industry best practices

Industry best practices for multi-cloud Kubernetes environments can help organizations optimize their deployments, improve security, and increase efficiency. Another benefit is new team members who are familiar with the industry best practices are already familiar and can have an accelerated learning curve. Some of the key best practices include:

- **Standardize tooling and processes**: Choose a consistent set of tools and processes for managing your Kubernetes environments across different cloud providers. This will simplify management, reduce the learning curve, and streamline workflows.
- **Centralized monitoring and logging**: Implement a centralized monitoring and logging solution to collect metrics and logs from all your Kubernetes clusters across different cloud providers. This will enable better observability, faster troubleshooting, and easier analysis of the overall system health.
- **Security best practices**: Implement robust security measures across all Kubernetes clusters, including **role-based access control (RBAC)** which uses SSO sources which are tied to company directories, network policies, and encryption for data at rest and in transit. Regularly scan container images for vulnerabilities and use a trusted registry.
- **Disaster recovery and business continuity planning**: Develop a comprehensive disaster recovery and business continuity plan that addresses the risks associated with multi-cloud environments. Regularly test the plan to ensure that it works as expected.
- **Data sovereignty and compliance**: Ensure that your multi-cloud Kubernetes deployment complies with data sovereignty and regulatory requirements in each region where your clusters are deployed. This may include storing data in specific regions or implementing specific security measures.

- **Optimize resource usage**: Use auto-scaling and right-sizing strategies to ensure that your Kubernetes clusters are using resources efficiently. Monitor resource usage to identify opportunities for optimization and cost savings.
- **Multi-cloud networking**: Implement a secure and efficient networking strategy that enables communication between Kubernetes clusters across different cloud providers. This may include using **VPNs**, dedicated interconnects, or other secure connectivity options.
- **CI/CD**: Implement a CI/CD pipeline that supports deployments across multiple Kubernetes clusters and cloud providers. This will help ensure consistent application releases and reduce the risk of errors during deployment.
- **Documentation and knowledge sharing**: Maintain thorough documentation of your multi-cloud Kubernetes environment, including architecture diagrams, deployment processes, and troubleshooting guides. Share this information with team members to ensure smooth collaboration and effective management of the environment.

By following these industry best practices, organizations can optimize their multi-cloud Kubernetes deployments, improve security, and maximize efficiency.

Conclusion

In conclusion, this chapter has provided a comprehensive overview of multi-cloud Kubernetes, exploring its evolution, advantages and disadvantages, key considerations, and best practices. We have examined hypothetical use cases and real-world examples to demonstrate the value of adopting a multi-cloud Kubernetes strategy. By understanding the importance of thorough planning, addressing potential challenges, and leveraging best practices, organizations can effectively deploy and manage Kubernetes environments across multiple cloud providers. This approach offers increased flexibility, resilience, and scalability, ultimately empowering businesses to stay competitive and agile in an ever-changing technological landscape.

In the next chapter, we will dive into some common scenarios of multi-cloud deployments, specifically stateful and stateless based Kubernetes clusters. We will also look at some use cases along with their advantages and challenges.

Join our book's Discord space

Join the book's Discord Workspace for Latest updates, Offers, Tech happenings around the world, New Release and Sessions with the Authors:

https://discord.bpbonline.com

CHAPTER 3
Scenarios of Multi-Cloud Deployment

Introduction

In this chapter, we will explore the practical and philosophical differences between two primary types of multi-cloud Kubernetes deployments: stateful application-based clusters and stateless application-based clusters. By examining the unique characteristics, requirements, and approaches associated with each type of deployment, we aim to provide readers with a deeper understanding of the various ways applications can be managed in a multi-cloud Kubernetes environment.

Structure

The chapter covers the following topics:
- Stateful and stateless applications
- Difference between the two types of clusters
- Stateful application-based clusters
- Use cases where stateful clusters are best suited
- Stateless application-based clusters
- Use cases where stateless clusters best suited
- Use cases

- Advantages
- Challenges

Objectives

By the end of this chapter, our goal is to enable readers to understand the practical and philosophical differences between stateful and stateless applications in a multi-cloud Kubernetes environment. Readers should be able to grasp the unique challenges and considerations associated with each deployment type, as well as the advantages and trade-offs of these approaches. Armed with this knowledge, readers will be better prepared to make informed decisions when designing and implementing multi-cloud Kubernetes deployments that align with their organization's specific needs and requirements.

Stateful and stateless applications

Stateful and stateless applications refer to two different approaches to handling and storing data in the context of software applications. Let us go over them both.

Stateful applications

Stateful applications are those that maintain and rely on state information to function correctly. The state can include data such as user sessions, transactions, or any other information that persists across multiple requests or interactions. In a stateful application, the application server retains this state data between requests, either in memory or some persistent storage.

Examples of stateful applications include databases, message brokers, or web applications that need to keep track of user sessions, shopping carts, or other user-specific data.

Stateless applications

Stateless applications, on the other hand, do not store any state information or depend on stored data between requests. They process each request independently, relying solely on the input provided within the request, without considering any previous interactions. As a result, stateless applications can be easily scaled horizontally by adding more replicas to handle increased workloads, as there is no state to synchronize between instances.

Examples of stateless applications include RESTful APIs, static web servers, or any other application that processes requests without needing to maintain any user-specific data or state information.

In summary, the primary difference between stateful and stateless applications lies in how they handle and store data. Stateful applications maintain state information, which is crucial to their functionality, while stateless applications do not rely on any stored state and process requests independently. Understanding these differences is essential when designing and implementing software applications to align with specific needs and requirements.

Differences between two types of clusters

In the context of multi-cloud Kubernetes deployments, the practical and philosophical differences between stateful and stateless application-based clusters can be summarized as follows.

Data persistence and management

Stateful applications maintain and rely on state information, necessitating careful planning and management of data persistence, replication, and synchronization across multiple cloud providers. In a stateful application-based cluster, Kubernetes uses a variety of APIs such as StatefulSets, Deployments, DaemonSet or Pods along with persistent volumes to ensure data persistence and unique identities for workloads. Managing data in a multi-cloud environment becomes more complex, requiring cross-cloud data replication and synchronization strategies to maintain consistency and durability of the data.

Stateless applications, on the other hand, do not store any state information or depend on stored data. They can be deployed and managed more easily across multiple cloud providers without the need for complex data management techniques since there is no state to synchronize between clouds.

Scalability

Scaling stateful applications in a multi-cloud environment can be challenging due to the need to maintain consistency and data integrity during the scaling process. Techniques like sharding, partitioning, and consistent hashing must be employed to distribute data and workload across multiple nodes and cloud providers while preserving data consistency.

Stateless applications, by contrast, are more straightforward to scale in a multi-cloud context as they do not depend on stored state. They can be easily scaled horizontally by adding more replicas to handle the increased workload across multiple cloud providers, without the need for complex data management techniques.

High availability and fault tolerance

In a multi-cloud environment, stateful applications demand more sophisticated strategies to ensure high availability and fault tolerance. This includes managing data replication across different cloud providers, implementing backup and disaster recovery plans, and addressing network partitions and other distributed system challenges.

Stateless applications, due to their lack of reliance on state information, can be more resilient to failures and easier to recover. New replicas can be quickly spun up to replace failed instances without the need for complex recovery processes, making stateless applications inherently more fault-tolerant and highly available in a multi-cloud environment.

Deployment and maintenance complexity

Managing stateful applications in a multi-cloud Kubernetes environment entails more effort and expertise due to the complexities related to data management, replication, and consistency. This can lead to increased deployment and maintenance costs and a steeper learning curve for development and operations teams.

Stateless applications are generally easier to deploy and maintain, as they do not have dependencies on stored state and are less prone to data-related issues. This makes stateless applications more attractive for organizations seeking to minimize complexity and operational overhead in their multi-cloud Kubernetes deployments.

In summary, within the context of multi-cloud Kubernetes deployments, the primary differences between stateful and stateless application-based clusters lie in data persistence and management, scalability, high availability, and deployment complexity. Understanding these differences is crucial for organizations when designing and implementing multi-cloud Kubernetes deployments that align with their specific needs and requirements.

Refer to the following *Table 3.1*:

Feature	Stateful Kubernetes Clusters	Stateless Kubernetes Clusters
State persistence	State is maintained and preserved	State is not maintained or preserved
Data storage	**Persistent Volumes (PV)**	Ephemeral storage or external storage services
Scaling	Manual or autoscaling with careful management of state	Easy horizontal scaling without state management concerns
Failover and recovery	Failover requires restoring state from backup or replica	Failover does not require restoring state; new replicas can start processing requests immediately
Dependency on previous interactions	Application behavior depends on the state from previous interactions	Application behavior is independent of previous interactions

Table 3.1: Features of stateful and stateless Kubernetes clusters

Stateful application-based clusters

In this section of the chapter, we will delve deeper into stateful application-based clusters, exploring the characteristics of stateful applications and how they interact with some of the fundamental Kubernetes features. Additionally, we will discuss the increased complexity of deploying stateful applications to multi-cloud Kubernetes clusters.

Stateful applications

A stateful application is one that stores and relies on state information, which can include user sessions, transactions, or any other data that persists across multiple requests. This state information is critical to the application's functionality and must be managed carefully to ensure consistency, availability, and durability. Refer to the following figure:

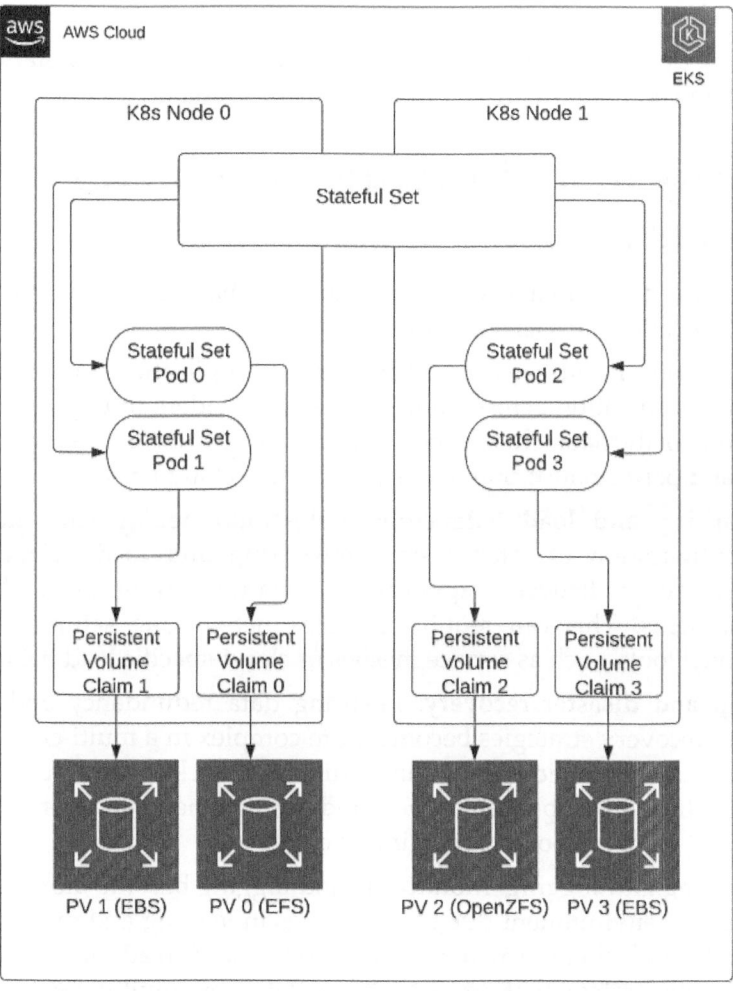

Figure 3.1: Stateful application running as a stateful set

Kubernetes and stateful applications

Kubernetes offers several features that support the deployment and management of stateful applications, including the following:

- **StatefulSets**: StatefulSets are a higher-level abstraction over pods designed specifically for stateful applications. They provide guarantees on the ordering and uniqueness of pods, ensuring that each pod gets a stable hostname based on a unique index (for example, web-0, web-1, web-2).

- **Persistent Volumes and Persistent Volume Claims**: To store data persistently, Kubernetes uses the concepts of **PV** and **Persistent Volume Claims (PVC)**. PVs represent physical storage resources in a cluster, while PVCs are requests for storage resources by users. StatefulSets can use PVC templates to dynamically provision and bind PVs for each pod, ensuring data persistence across pod restarts and rescheduling.

- **Services**: Services are Kubernetes objects that can provide a static IP address, a DNS name both internally or externally for workloads.

Complexity in multi-cloud Kubernetes deployments

Deploying stateful applications in a multi-cloud Kubernetes environment introduces additional complexities, such as the following:

- **Data replication and synchronization**: Stateful applications require data replication and synchronization across multiple cloud providers to ensure consistency and durability of the data. This can be challenging due to differences in storage services, APIs, and performance characteristics between cloud providers.

- **Networking and load balancing**: Multi-cloud deployments may necessitate the configuration of inter-cluster networking and load balancing to enable communication between application instances running in different cloud environments. This can require complex networking setups and the use of additional tools, such as service meshes or cloud-specific load balancers.

- **Backup and disaster recovery**: Ensuring data redundancy and implementing disaster recovery strategies become more complex in a multi-cloud environment, as each cloud provider may have different backup and recovery solutions. Organizations need to carefully plan and manage their backup strategies to ensure consistent recovery points and minimize data loss.

- **Monitoring and logging**: Monitoring and logging become more challenging in a multi-cloud environment due to differences in logging and monitoring solutions offered by each cloud provider. Organizations need to adopt a unified monitoring and logging strategy that can consolidate data from multiple sources and provide a holistic view of the application's performance and health across all cloud environments.

Use cases where stateful clusters are best suited

Stateful application-based clusters are best suited for business use cases that require the storage and management of state information across multiple requests or interactions. Some common examples are as follows:

- **E-commerce platforms**: E-commerce platforms need to maintain user-specific information such as shopping carts, wish lists, and user profiles. A stateful application-based cluster can store and manage this data, ensuring seamless user experiences across various devices and sessions.
- **Customer relationship management (CRM) systems**: CRM systems manage customer information, interactions, and sales processes. Stateful application-based clusters can store and manage customer data, allowing sales representatives to view and update customer profiles, track interactions, and manage leads and opportunities.
- **Online gaming**: Online gaming platforms often require stateful application-based clusters to manage user accounts, game state, and in-game transactions. These clusters can store game progress, player rankings, and virtual currencies, ensuring a consistent experience for users across sessions and devices.
- **Content Management Systems (CSM)**: CMS platforms manage the creation, editing, and publishing of digital content. Stateful application-based clusters can store and manage content, user data, and access permissions, enabling authors and editors to collaborate on content creation and management.
- **Financial services**: Financial services applications, such as online banking, trading platforms, and payment systems, often require stateful application-based clusters to manage user accounts, transactions, and account balances. These clusters ensure that financial data remains consistent and secure across multiple interactions and requests.
- **IoT and real-time analytics**: In IoT and real-time analytics use cases, stateful application-based clusters can manage the continuous ingestion, processing, and storage of streaming data from sensors and devices. These clusters can maintain the state of various data processing pipelines and ensure that data is consistently processed and analyzed in real-time.
- **Social networking platforms**: Social networking platforms require stateful application-based clusters to store and manage user profiles, relationships, and interactions. These clusters can maintain user connections, messages, and

multimedia content, ensuring a consistent user experience across various devices and sessions.

In conclusion, stateful application-based clusters are best suited for use cases where maintaining and managing state information is crucial to the application's functionality and user experience. These clusters provide the necessary infrastructure and features to store and manage state data effectively and efficiently.

Stateless application-based clusters

In this section, we will focus on stateless application-based clusters, highlighting their unique characteristics and their implications on Kubernetes features. We will also discuss the increased complexity of stateless applications when deployed across multi-cloud Kubernetes clusters. Stateless applications are those that do not store or depend on any state information between requests or interactions. They process each request independently, based solely on the input provided within the request. Stateless applications can be easily scaled horizontally by adding more instances to handle increased workloads, as there is no state to synchronize between instances.

Refer to the following figure:

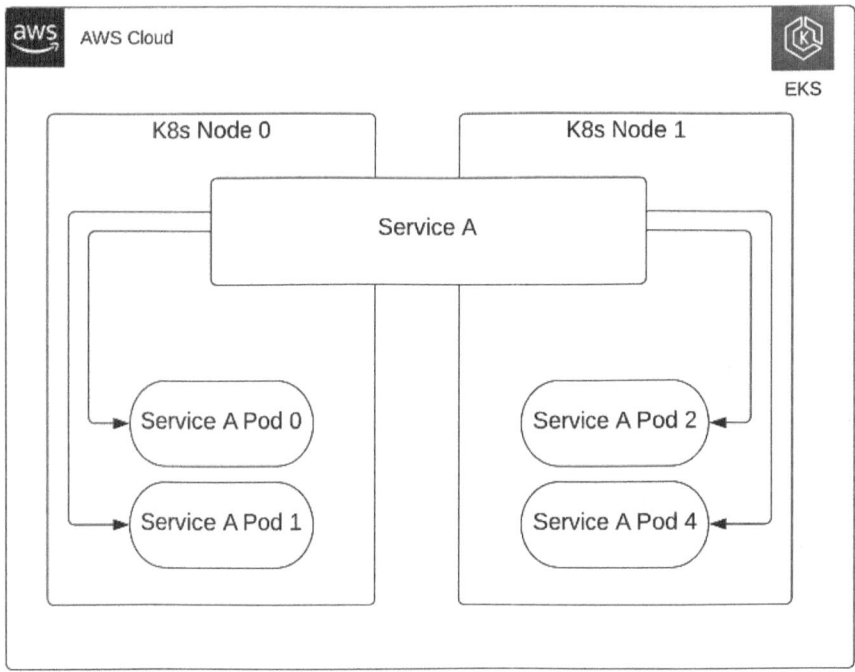

Figure 3.2: Stateless application running as a StatefulSet

Implications

Stateless applications have several implications on basic Kubernetes features, such as the following:

- **Scalability**: Stateless applications are easier to scale in Kubernetes, as they can be replicated across multiple instances without the need to maintain state synchronization. This enables efficient load balancing and distribution of traffic among different instances.
- **High availability**: Stateless applications are inherently more resilient to failures, as each request is independent of any previous interactions. This allows Kubernetes to restart failed instances or replace them with new ones without impacting the application's functionality or user experience.
- **Simplified deployment and management**: Stateless applications generally have simpler deployment and management processes, as there is no state data to manage or migrate. Kubernetes can easily update and roll back stateless applications without worrying about data consistency or integrity.
- **Resource efficiency**: Stateless applications can efficiently utilize resources in a Kubernetes cluster, as they do not require the same level of storage or data management capabilities as stateful applications. This can lead to cost savings and better resource utilization.

Complexities

However, deploying stateless applications in a multi-cloud Kubernetes environment introduces additional complexities, such as the following:

- **Data consistency**: Although stateless applications do not store state themselves, they may still rely on external stateful services like databases. Ensuring data consistency across multiple cloud providers can be challenging, requiring techniques such as data replication, synchronization, or global load balancing.
- **Networking and security**: Deploying stateless applications across multiple cloud providers may necessitate complex networking configurations and security policies. Organizations must ensure seamless connectivity, secure data transmission, and compliance with various cloud providers' security requirements.
- **Monitoring and logging**: Managing logs and monitoring metrics from stateless applications deployed in a multi-cloud environment can be complex. Organizations must consolidate and analyze logs and metrics from multiple sources and cloud providers to gain a holistic understanding of their applications' performance and health.
- **Cost management**: Deploying stateless applications across multiple cloud providers can result in increased infrastructure costs. Organizations must carefully

manage and optimize their resource usage across different cloud environments to maintain cost efficiency.

In conclusion, stateless application-based clusters offer several advantages in terms of scalability, high availability, and simplified deployment in Kubernetes. However, deploying stateless applications in a multi-cloud Kubernetes environment introduces additional complexities related to data consistency, networking, security, monitoring, and cost management. Addressing these challenges requires careful planning and implementation of best practices and tools designed for multi-cloud Kubernetes deployments.

Use cases where stateless clusters are best suited

Stateless application-based clusters are best suited for business use cases where maintaining state information is not critical, and the applications can process each request independently. Some common examples include the following:

- **Web servers**: Stateless application-based clusters are ideal for serving static or dynamic web content, as they can efficiently handle incoming requests without relying on any previous interactions. This enables seamless scaling and high availability.
- **RESTful APIs**: Stateless APIs enable developers to build scalable and resilient backend services by processing each API request independently. This allows for easy horizontal scaling and supports load balancing across multiple instances in a Kubernetes cluster.
- **Microservices architectures**: Stateless microservices can be easily scaled, updated, and replaced without affecting other services within the system. This allows for better resource utilization, easier deployment, and more resilient application architecture.
- **Serverless functions**: Serverless computing platforms, like AWS Lambda or Azure Functions, are inherently stateless, allowing developers to create event-driven applications that automatically scale with the number of requests. Stateless application-based clusters can help manage and deploy these serverless functions in a Kubernetes environment.
- **Content Delivery Networks (CDN)**: CDN distribute web content, such as images, videos, and scripts, to edge servers worldwide, ensuring fast content delivery and reduced latency. Stateless application-based clusters are well-suited for CDNs, as they can handle multiple requests independently and efficiently.
- **Load balancers and reverse proxies**: Load balancers and reverse proxies distribute incoming traffic across multiple backend services, ensuring optimal resource utilization and improved application performance. Stateless application-based clusters can easily scale and manage these networking components.

- **Batch data processing**: Stateless application-based clusters are ideal for batch data processing tasks, such as data transformation, aggregation, or analysis. Each task can be processed independently, allowing for efficient resource allocation and horizontal scaling.

In summary, stateless application-based clusters are best suited for use cases where state information is not crucial, and applications can process each request independently. These clusters enable efficient horizontal scaling, high availability, and simplified deployment and management, making them an excellent choice for various business scenarios.

Use cases

Refer to the following *Table 3.2*:

Application type	Stateful clusters	Stateless clusters
Database	Suitable for databases requiring data persistence	Not ideal for databases
Web server	Can be used for session-based applications	Suitable for RESTful APIs and static content
Cache	Suitable for caching solutions	Can be used for cache systems with no persistence
Message queue	Suitable for message brokers requiring persistence	Can be used for message brokers without persistence

Table 3.2: Use cases for clusters

Advantages

Refer to the following *Table 3.3:*

Feature	Stateful clusters	Stateless clusters
Data persistence	Data is preserved across restarts	Data is not preserved
Fault tolerance	Can recover state from backups/replicas	Quick recovery without restoring state

Table 3.3: Advantages of clusters

Challenges

Refer to the following *Table 3.4*:

Feature	Stateful clusters	Stateless clusters
Scaling	Can require careful state management	Easily horizontally scalable
Complexity	Higher due to state management	Lower complexity
Storage management	Requires proper management of PVs and PVCs	May require external storage services

Table 3.4: Challenges with clusters

Conclusion

In this chapter, we have explored the practical and philosophical differences between stateful and stateless application-based clusters in multi-cloud Kubernetes environments. We discussed the unique characteristics, advantages, and challenges associated with both stateful and stateless clusters, providing a comprehensive understanding of their implications in various business scenarios.

Stateful application-based clusters are better suited for applications that require persistent data storage, while stateless application-based clusters excel in scenarios where applications can process each request independently and without relying on previous interactions. We have also provided real-world use cases to demonstrate the applicability of each approach in different business contexts.

As we move forward, the next two chapters will talk about Kubernetes cluster design for both stateful and stateless clusters. We will provide more detailed examples, including actual stateful and stateless application code that can run on a Kubernetes cluster. These chapters will equip you with the knowledge and practical skills necessary to design, deploy, and manage stateful and stateless applications effectively in multi-cloud Kubernetes environments.

By understanding the differences and nuances between stateful and stateless application-based clusters, you will be better prepared to make informed decisions when architecting your multi-cloud Kubernetes solutions. Stay tuned as we dive deeper into the world of Kubernetes cluster design and uncover the secrets to building robust and scalable stateful and stateless applications in the cloud.

Chapter 4
Stateful Application Kubernetes Cluster Design

Introduction

In this chapter, we will dive into the architecture, design, and implementation of Kubernetes clusters specifically tailored to support stateful applications. This chapter will include concrete system designs and project plans, making it a valuable resource for those looking to kickstart their multi-cloud Kubernetes journey. We will examine a variety of use cases and delve into the many methods of storing data across multiple cloud providers. By providing practical, real-world examples and actionable insights, this chapter aims to equip you with the knowledge and tools needed to design and deploy stateful applications on multi-cloud Kubernetes clusters.

Structure

The chapter covers the following topics:
- Overview of stateful applications
- Architecture and design considerations for stateful application Kubernetes clusters
- Data storage strategies across multiple cloud providers
- Sharing real-time data across clusters in different cloud providers
- Best practices for stateful application design in Kubernetes

- Stateful application Kubernetes cluster design patterns
- Example of stateful application deployments on multi-cloud Kubernetes

Objectives

By the end of this chapter, readers will have a deep understanding of the key architectural and design considerations for building stateful application Kubernetes clusters in a multi-cloud environment. They will become familiar with the specific Kubernetes resources essential for stateful applications and learn how to properly use them. Readers will also explore various data storage strategies and options across multiple cloud providers, while being introduced to best practices and design patterns for stateful application deployments in Kubernetes. Furthermore, hands-on experience with concrete system designs and project plans will be provided, guiding readers through their own stateful application cluster implementation. Real-world use cases and examples of stateful application deployments on multi-cloud Kubernetes will be examined, offering valuable insights for readers to apply in their projects.

Overview of stateful applications

Stateful applications are those that maintain and rely on a persistent state between different requests or client interactions. The application's state includes any data that is stored and updated during its execution, allowing it to remember previous interactions, user preferences, or other relevant information. Stateful applications often require more complex management and architecture, as the state must be maintained, synchronized, and shared across multiple instances or nodes.

The key characteristic of stateful applications is their dependence on previously stored data to process new requests or provide meaningful responses. This means that stateful applications typically require a mechanism to store and manage the state, which can include in-memory storage, databases, or distributed storage systems.

Some common examples of stateful applications include:
- Online shopping carts, where the application remembers the items added by a user between different sessions.
- User authentication systems, where the application maintains session information to identify logged-in users and their access permissions.
- Online gaming platforms, where the application tracks user progress, scores, and game state.

Stateful applications offer several advantages in specific use cases:
- **Personalization:** By maintaining user data and preferences, stateful applications can provide personalized experiences tailored to individual users.

- **Optimized performance:** Stateful applications can cache data and state, reducing the need for repetitive queries or computations, thereby improving performance.
- **Complex interactions:** Stateful applications can support complex, multi-step interactions that require maintaining and updating the state throughout the process.

However, stateful applications also introduce challenges in a cloud-native context, such as:

- **Scalability:** Scaling stateful applications can be more complex, as new instances may need to access or synchronize the state with existing instances.
- **Availability:** Ensuring high availability for stateful applications can be challenging, as failures or disruptions might lead to loss or corruption of the state.
- **Data consistency:** Managing and maintaining data consistency across distributed stateful applications can be complicated, especially when dealing with concurrent updates or requests.

When architecting stateful applications, it is crucial to carefully consider the mechanisms for state management, synchronization, and data persistence, as well as the trade-offs between performance, scalability, and availability.

Considerations for stateful application clusters

In this section, we will explore the architecture and design considerations for stateful application Kubernetes clusters, specifically addressing the fact that storage options vary depending on the cloud provider. Designing a Kubernetes cluster that can effectively handle stateful applications requires careful planning and thoughtful decision-making. Here are some key factors to consider when creating an architecture for stateful applications on Kubernetes:

- **StatefulSets:** Kubernetes provides a resource called StatefulSet, which is specifically designed for managing stateful applications. StatefulSets ensure that each pod has a unique, stable hostname and maintain a stable network identity across rescheduling. This allows for ordered and graceful scaling, updates, and deletion, as well as persistent storage with **Persistent Volumes (PV)**.
- **Persistent storage:** Stateful applications often require persistent storage for their data. Since storage options vary depending on the cloud provider, it is essential to choose the right storage solution based on your application's performance, reliability, and scalability requirements. Kubernetes supports multiple storage options such as **Network-Attached Storage (NAS)**, block storage, object storage and distributed storage systems like Amazon EBS, Google Persistent Disk, or Azure Disk Storage. Evaluate each cloud provider's storage offerings and select the one that best fits your needs.

- **Data replication and backup**: Stateful applications need a strategy to ensure data consistency, replication, and backup. Depending on the storage solution and cloud provider, you may need to configure replication and backup policies to protect against data loss and ensure high availability. Some databases and storage systems provide built-in data replication and backup features, while others require manual or third-party solutions.
- **Network configuration**: Multi-cloud Kubernetes clusters often span different networks, making network configuration and communication essential. Consider how services will communicate across different cloud providers, and make use of ingress controllers, load balancers, and service meshes to maintain connectivity, performance, and security.
- **Disaster recovery and failover**: Design your cluster to handle failures and ensure high availability. Consider using multi-zone or multi-region deployments and configuring replica sets for critical services. Implement a disaster recovery plan that includes backup and restore procedures, as well as a failover strategy to maintain application availability during outages.
- **Monitoring and observability**: Keep an eye on the health and performance of your stateful application Kubernetes cluster using monitoring tools like Prometheus and Grafana. Track metrics, logs, and traces to ensure your application is running optimally and troubleshoot issues when they arise.
- **Security**: Secure your cluster by following best practices, such as the principle of least privilege, **Role-Based Access Control (RBAC)**, and proper network segmentation. Make use of Kubernetes-native security features, such as Network Policies and Pod Security Admission, often referred to as **PodSecurityPolicy (PSP)**, and utilize third-party tools to enforce security policies and monitor for vulnerabilities.

By considering these factors and applying best practices, you can design a robust, scalable, and reliable Kubernetes cluster that effectively supports stateful applications across multiple cloud providers. This will enable your organization to fully harness the benefits of a multi-cloud Kubernetes environment while minimizing potential risks and challenges.

Data storage strategies across multiple cloud providers

In the following section, we discuss various data storage strategies for stateful applications running on Kubernetes clusters across multiple cloud providers. Handling data storage in a multi-cloud environment can be challenging, as each cloud provider offers different storage solutions and services. The goal is to achieve a seamless and efficient way to store, manage, and access data across all clouds involved. Here are some strategies to consider:

- **Data replication and synchronization**: One approach to managing data across multiple cloud providers is to use data replication and synchronization tools. These tools ensure that data is consistently available across all clouds, allowing your application to read and write data regardless of the underlying cloud provider. Some examples of such tools are database replication technologies (for example, MySQL replication, PostgreSQL replication) or file synchronization tools, such as, Rsync.
- **Hybrid cloud storage**: Hybrid cloud storage solutions, such as NetApp Cloud Volumes, can help you manage data across multiple cloud providers by providing a single, unified storage layer that works seamlessly across all clouds. This approach abstracts the underlying cloud storage services and provides a consistent interface for your applications, simplifying data management and migration.
- **Multi-cloud storage gateways**: Multi-cloud storage gateways, such as CloudianHyperStore or AverevFXT, are software-defined storage solutions that provide a single, unified interface for accessing and managing data across multiple cloud providers. They can be deployed as a service or a virtual appliance within your Kubernetes cluster and can be integrated with various cloud provider storage services, such as Amazon S3, Google Cloud Storage, or Azure Blob Storage.
- **Cloud provider storage abstraction libraries**: Another approach to handling data storage across multiple cloud providers is to use cloud provider storage abstraction libraries, such as Apache Libcloud or HashiCorp's Terraform. These libraries provide a consistent API for interacting with different cloud provider storage services, making it easier to develop applications that work across multiple clouds.
- **Application-level sharding**: In some cases, it might be beneficial to split your application data across multiple cloud providers based on specific requirements or characteristics, such as latency, cost, or regulatory compliance. This can be achieved by implementing application-level sharding, where each shard is responsible for a specific subset of data and is hosted on a different cloud provider.
- **Cross-cloud backups**: Ensuring data durability and disaster recovery is crucial in a multi-cloud environment. Regularly backing up your data to another cloud provider can provide an additional layer of protection in case of data loss or service outages in your primary cloud provider.

When designing your data storage strategy for a multi-cloud Kubernetes environment, it is essential to consider factors such as data consistency, performance, cost, and vendor lock-in. By evaluating these factors and choosing the appropriate strategy for your specific use case, you can ensure that your stateful applications run smoothly and efficiently across multiple cloud providers.

Sharing real-time data across clusters in different cloud providers

The next topic we cover is sharing real-time data across Kubernetes clusters in different cloud providers. When running stateful applications in a multi-cloud Kubernetes environment, it is crucial to ensure that data is available, consistent, and easily accessible across all clusters. Here, we discuss some strategies and techniques to achieve real-time data sharing between Kubernetes clusters in different cloud providers:

- **Distributed databases**: Using a distributed database can help maintain data consistency and availability across multiple clusters. Distributed databases, such as Cassandra, CockroachDB, or Amazon Aurora, are designed to replicate and synchronize data across multiple nodes in different locations. This ensures that your application can read and write data in real-time, regardless of which cluster or cloud provider it is running on.
- **Global load balancers**: Global load balancers, such as Google Cloud's Global Load Balancing or AWS Global Accelerator, can help distribute traffic across multiple clusters in different cloud providers. By routing traffic based on factors such as latency or user location, global load balancers can ensure that your application's data is accessed efficiently and consistently across all clusters.
- **Message queues and streaming platforms**: Message queues, such as Apache Kafka or RabbitMQ, and streaming platforms, such as Apache Flink or Google Cloud Dataflow, can help share real-time data between applications running on different clusters or cloud providers. These tools provide a scalable and fault-tolerant way to transmit data between applications, allowing them to communicate and share data in real-time.
- **API gateways and service mesh**: API gateways, such as Kong or Amazon API Gateway, and service mesh solutions, such as Istio or Linkerd, can help facilitate communication between microservices running on different clusters or cloud providers. By providing a consistent and secure way for services to communicate, these tools can enable real-time data sharing between applications, regardless of their underlying cloud provider.
- **Custom synchronization solutions**: In some cases, you may need to develop custom synchronization solutions to handle real-time data sharing between clusters in different cloud providers. This could involve using replication and synchronization tools, such as database replication technologies or file synchronization tools, to ensure that data is consistently available across all clusters.

When implementing real-time data sharing between Kubernetes clusters in different cloud providers, it is important to consider factors such as latency, data consistency, and security. By choosing the appropriate strategy and tools for your specific use case, you can ensure that your stateful applications can access and share data in real-time, regardless of the underlying cloud provider.

Best practices for stateful application design

Designing stateful applications for Kubernetes requires thoughtful planning and adherence to best practices to ensure scalability, reliability, and maintainability. Here are some best practices for stateful application design in Kubernetes:

- **Leverage StatefulSets**: StatefulSets are a Kubernetes resource specifically designed to manage stateful applications. They provide guarantees around the ordering and uniqueness of pods, making it easier to manage applications that require persistent storage or stable network identities.

- **Use PV and Persistent Volume Claims (PVC)**: PV and PVC are Kubernetes resources that enable the provisioning and management of persistent storage for stateful applications. By using PVs and PVCs, you can abstract the underlying storage infrastructure and ensure that your application data is stored in a reliable and durable manner.

- **Design for horizontal scaling**: Stateful applications should be designed to scale horizontally by adding more instances to handle increased load. This involves architecting your application to support data partitioning and replication across multiple instances.

- **Graceful handling of failures**: Ensure your application can handle node, pod, or storage failures gracefully. Implement appropriate retry mechanisms and consider using circuit breakers or other resilience patterns to minimize the impact of failures on your application's functionality and performance.

- **Implement health checks**: Kubernetes supports liveness and readiness probes to determine the health of your application. Implement these health checks to help Kubernetes automatically restart or reschedule unhealthy instances, ensuring the continued availability of your application.

- **Secure your application**: Implement proper authentication, authorization, and encryption mechanisms to protect your application's data. This includes using Kubernetes Secrets to store sensitive information, implementing RBAC for application components, and using encryption for data at rest and in transit.

- **Monitor and log**: Implement comprehensive monitoring and logging for your stateful application to quickly identify and troubleshoot issues. Leverage Kubernetes-native tools like Prometheus for monitoring and Splunk, Sumo Logic or Datadog for log aggregation. There are many other third-party tools to gain insights into your application's performance and behavior, your requirements will guide you to your solution.

- **Optimize resource utilization**: Configure your stateful application to use resources efficiently. Set appropriate resource requests and limits for CPU, memory, and storage to help Kubernetes manage the allocation of resources effectively.

- **Backup and disaster recovery**: Plan and implement a robust backup and disaster recovery strategy for your stateful application. Regularly backup your application data and test your recovery process to ensure minimal downtime and data loss in the event of a disaster.

- **Automate deployment and management**: Use tools like Helm or Kustomize to automate the deployment and management of your stateful application in Kubernetes. This helps to streamline the deployment process and ensures consistency across different environments.

By adhering to these best practices, you can design stateful applications that are scalable, reliable, and maintainable, ensuring smooth operation in a Kubernetes environment.

Stateful application Kubernetes cluster design patterns

Stateful application Kubernetes cluster design patterns provide solutions for common challenges when running stateful applications in a Kubernetes environment. These patterns focus on managing state, ensuring data consistency, and maintaining high availability. Here are some design patterns for stateful applications in Kubernetes:

- **StatefulSet pattern**: Use StatefulSets to manage stateful applications, as they provide guarantees about the ordering and uniqueness of pods. StatefulSets maintain a stable hostname for each pod, ensuring a consistent network identity for stateful applications.
- **Persistent storage pattern**: Leverage PV and PVC to provide a consistent storage interface for your applications. PVs and PVCs abstract the underlying storage infrastructure, allowing applications to work with storage resources without being tied to a specific cloud provider or storage system.
- **Sharding pattern**: Distribute data across multiple instances or nodes in your cluster to improve overall performance and handle increased load. Sharding can be implemented through consistent hashing, range-based partitioning, or other techniques to distribute data evenly and minimize hotspots.
- **Replication pattern**: Implement replication to maintain multiple copies of your data across different nodes, zones, or even cloud providers. This helps ensure high availability and fault tolerance, enabling your application to continue running even if some nodes or regions experience downtime.
- **Active-standby pattern**: Deploy an active-standby setup where one instance of the application is actively serving traffic (active) while another instance is on standby, ready to take over if the active instance fails. This pattern ensures high availability and seamless failover in case of node or application failure.
- **Active-active pattern**: Utilize an active-active configuration where multiple instances of the application are actively serving traffic and synchronizing their state. This pattern provides improved performance and load balancing, as well as increased fault tolerance.

- **Backup and recovery pattern**: Implement periodic backups and a disaster recovery strategy for your stateful applications. This may include taking snapshots, creating offsite backups, or replicating data to another cluster, ensuring that your application can recover from data loss or corruption.
- **Multi-zone or multi-cloud pattern**: Deploy your stateful application across multiple zones, regions, or cloud providers to ensure high availability and fault tolerance. This pattern helps minimize the impact of failures in a single zone, region, or cloud provider, reducing the risk of downtime.
- **Caching pattern**: Use caching to improve the performance of your stateful application by temporarily storing frequently accessed data in memory. This reduces the load on your primary data store and helps decrease response times for end-users.
- **Leader election pattern**: Implement a leader election mechanism for distributed applications that require coordination among multiple instances. This pattern ensures that only one instance is responsible for a specific task, preventing conflicts and ensuring consistency across the application.

By applying these design patterns to your stateful application Kubernetes cluster, you can effectively manage state, ensure data consistency, and maintain high availability for your applications.

Example of stateful application deployments

Let us now look at a common stateful application using what we have covered so far. This detailed example will provide readers with practical insights and a solid foundation for designing and implementing their own multi-cloud stateful applications.

For this exercise, we will use Redis as our example stateful application to deploy on two Kubernetes clusters, one in Amazon **Elastic Kubernetes Service (EKS)** and the other in **Google Kubernetes Engine (GKE)**. Redis is an open-source, in-memory data structure store that can be used as a database, cache, and message broker. It is widely used for its high performance, scalability, and flexibility in handling various data types. These two deployments will form a Redis Cluster and will keep its logs on a PV. We are going to go old school here and use Kubernetes YAML Manifests and be applying them with KubeCTL. More complex methods like Terraform and Helm will come later in the book.

Challenges of deploying Redis

Deploying Redis in a multi-cloud Kubernetes environment presents specific challenges that need to be addressed for a successful implementation. These challenges include the following:

- **Data consistency and replication**: Ensuring that the data stored in Redis instances across both cloud providers remains consistent and synchronized. This may involve setting up master-slave replication or using Redis Cluster for automatic sharding and high availability.
- **Networking and latency**: Configuring secure and reliable network connectivity between the two Kubernetes clusters, while minimizing latency for data access and replication.
- **Persistent storage**: Managing the storage backend for Redis instances to ensure data durability and resilience in case of node failures. This will involve working with PVs and PVCs in both EKS and GKE environments.
- **High availability and failover**: Ensuring that Redis instances are highly available and can automatically recover from failures, with minimal impact on the application's performance and user experience.
- **Monitoring and management**: Setting up monitoring and management tools to gain insights into Redis performance, resource usage, and potential issues across both Kubernetes clusters.

By addressing these challenges, we can successfully deploy a Redis stateful application in a multi-cloud Kubernetes environment, taking advantage of the benefits of both EKS and GKE.

Multi-cloud Redis system design

In this section, we will dive into the system design for our Redis deployment in a multi-cloud Kubernetes environment, focusing on four main components: Architecture Overview, Storage Options, Data Replication and Synchronization, and Networking. We will consider the differences between EKS and GKE for each of these components.

Architecture overview

A typical Redis deployment in a multi-cloud Kubernetes environment consists of Redis instances running in both EKS and GKE clusters. Depending on the desired level of high availability and fault tolerance, you can choose between Redis master-slave replication or a Redis Cluster setup.

For EKS and GKE, you will need to create Kubernetes manifests for the Redis instances, including Deployments, StatefulSets, Services, and ConfigMaps. You should also configure the necessary resources, such as CPU, memory, and storage for the Redis instances.

Refer to the following figure:

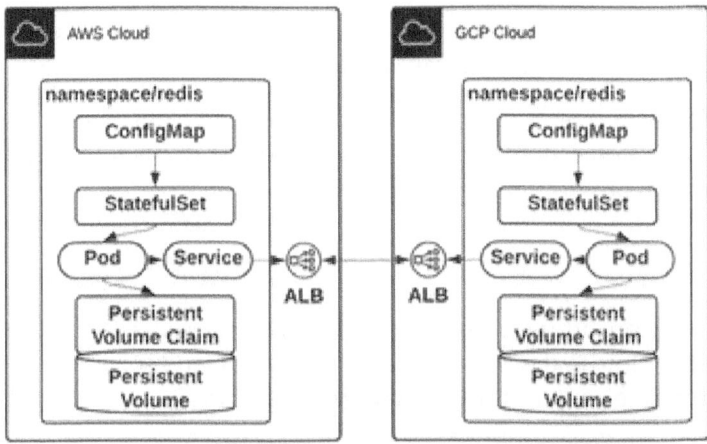

Figure 4.1: Multi-Cloud Redis Deployment on EKS and GKE

Storage options

To ensure data durability in both EKS and GKE environments, you need to configure PVs and PVCs for Redis instances. For EKS, you can use Amazon EBS volumes, while for GKE, you can use Google Persistent Disks. You will also need to configure storage classes and access modes based on the desired performance and reliability requirements.

Data replication and synchronization

To maintain data consistency and availability across both clusters, you can set up Redis master-slave replication or use a Redis Cluster. With master-slave replication, one Redis instance serves as the master, and the others act as read-only replicas. In a Redis Cluster, data is automatically partitioned across multiple instances for high availability and fault tolerance. Ensure that the necessary configurations, such as the replication factor, are set correctly for your desired level of data redundancy.

Networking

Establishing secure and reliable network connectivity between the EKS and GKE clusters is essential for data replication and synchronization. For EKS (AWS), you can use AWS Transit Gateway, AWS Direct Connect, or set up a Site-to-Site VPN connection with Google Cloud.

For GKE (Google Cloud), you can use Google Cloud VPN or VPC Network Peering with AWS. Additionally, you should configure Kubernetes Services, such as LoadBalancer or NodePort, to expose Redis instances within and outside the cluster. Be mindful of minimizing latency for data access and replication.

By thoroughly addressing each of these components and considering the differences between EKS and GKE, you can design and deploy a robust Redis stateful application in a multi-cloud Kubernetes environment.

Project plan

Here is a great project plan outline you can use for deploying Redis as a stateful application on both EKS and GKE. Note that every company is different, and so, use these as guidelines and fill in the appropriate content for your business use case.

Project name

Multi-cloud Redis Deployment on EKS and GKE

Project scope

The project aims to deploy Redis as a stateful application on existing Kubernetes clusters in both EKS and GKE environments. This deployment will ensure high availability, fault tolerance, and seamless data synchronization between the two cloud providers.

Objectives

The objectives of this project are:
- Design and implement a robust Redis deployment architecture for both EKS and GKE.
- Ensure data durability through proper storage configuration.
- Establish secure and reliable network connectivity between clusters.
- Maintain data consistency and availability through data replication and synchronization.
- Test and validate the deployment for functionality and performance.

Key deliverables

The key deliverables of this project are:
- Architecture design documentation.
- Kubernetes manifests for Redis deployment on EKS and GKE.
- Network connectivity configuration and documentation.

- Test plans and test results.
- Deployment and maintenance guides.

Project stakeholders and roles

The projects stakeholders and roles are as follows:

- **Project manager:** Oversees the project, ensures milestones are met, and manages resources.
- **Cloud architect:** Designs the architecture and ensures best practices for multi-cloud deployment.
- **Kubernetes engineer:** Implements and deploys the Redis instances on EKS and GKE.
- **Network engineer:** Establishes and maintains network connectivity between EKS and GKE.
- **QA engineer:** Develops and executes test plans to validate the deployment.

Project timeline

The project timeline is as follows:

- **Planning (1 week):** Define project objectives, scope, and deliverables. Identify stakeholders and their roles.
- **Design (2 weeks):** Design the architecture and networking for Redis deployment on EKS and GKE.
- **Implementation (3 weeks):** Develop Kubernetes manifests, configure storage, and establish network connectivity.
- **Testing (2 weeks):** Execute test plans, validate functionality, and address any issues found.
- **Deployment (1 week):** Deploy the final Redis instances on EKS and GKE and provide documentation.

Resources

The resources are as follows:

- **Personnel:** Project Manager, Cloud Architect, Kubernetes Engineer, Network Engineer, QA Engineer
- **Software:** Kubernetes, Redis, KubeCTL, AWS CLI, Google Cloud SDK

Risk management plan

The various risks and their management are as follows:

- **Risk**: Unplanned downtime or data loss during deployment
  - **Mitigation**: Implement proper backup and recovery strategies and perform thorough testing before deployment.
- **Risk**: Inadequate network connectivity between EKS and GKE
  - **Mitigation**: Monitor network performance and address any latency or connectivity issues proactively.
- **Risk**: Insufficient resources for the project
  - **Mitigation**: Regularly assess resource allocation and adjust as needed to ensure project success.
- **Risk**: Changes in cloud provider features or APIs
  - **Mitigation**: Stay up-to-date with cloud provider documentation and adapt the deployment as necessary.

Architectural diagram

Refer to the following figure:

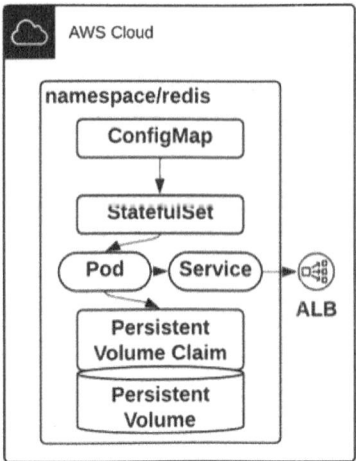

Figure 4.2: Architectural diagram of a Redis Service

Implementation steps for deploying Redis

Here, we finally get to build something! For this exercise, we are to crawl before we walk or run. We are going to go through the steps of deploying Redis on AWS EKS. In the coming chapters, we will build on this and add in the additional complexity of multi-cloud.

Prerequisites

The prerequisites are as follows:

- Ensure the Kubernetes cluster is up and running in EKS.
- Ensure the cluster is running in an isolated and protected environment. This cluster should have no access to any other resources, even development ones.
- Ensure the clusters has been set up for both public and private endpoint access.
- Install kubectl, AWS CLI, and Google Cloud SDK on your local machine.
- Configure access to the EKS using the respective CLI tools.
- On EKS, you must configure the EBS CSI driver for working PersistentVolumes.

Versions

This example assumes Kubernetes v1.26. We will be creating the following resources to deploy Redis:

- Namespace
- ConfigMap for Redis configuration
- StorageClass for EBS
- StatefulSet with the PVC
- Service to expose Redisexternally

We will first create the files, and then apply them after in order. We will go into more detail on each resource before we create them.

Base configuration

A Kubernetes Namespace is a logical partition or grouping within a Kubernetes cluster that allows you to isolate and manage resources more effectively. Namespaces provide a scope for resource names, enabling you to have multiple instances of the same resource in different namespaces without conflicts. They can be used to divide cluster resources among multiple users, teams, or projects, and they help to maintain a clean and organized environment. In addition to resource separation, namespaces can be used to apply resource quotas and network policies, further enhancing resource management, access control, and security within a cluster. The YAML code to create a namespace is as follows:

namespace.yaml

apiVersion: v1

kind: Namespace

metadata:

  name: redis

A Kubernetes ConfigMap is a resource used to store non-confidential configuration data in key-value pairs. It allows you to separate configuration information from the container image and the application code, making it easier to manage and update configurations independently. ConfigMaps enable you to decouple environment-specific configurations from your applications, which simplifies application deployment and scaling. Containers can consume ConfigMaps as environment variables, command-line arguments, or by mounting them as files in a volume. This provides a centralized and consistent way to manage configurations across multiple components in a Kubernetes cluster, making it more maintainable and flexible. The YAML code to create a configmap is as follows:

configmap.yaml

apiVersion: v1

kind: ConfigMap

metadata:

  name: redis-config

  namespace: redis

data: null

redis.conf: |

  bind 0.0.0.0

  protected-mode yes

  port 6379

  tcp-backlog 511

  timeout 0

  tcp-keepalive 300

  daemonize no

  supervised no

  pidfile /var/run/redis.pid

```
loglevel notice
logfile ""
databases 16
```

Storage class

A Kubernetes StorageClass is a resource that defines the storage provisioner, parameters, and other configuration details for creating PVs in a Kubernetes cluster. It allows administrators to define different classes of storage with varying performance, cost, and other characteristics, which can be requested by users through PVCs. StorageClass streamlines the dynamic provisioning of storage in a cluster, enabling users to obtain storage without needing to know the underlying infrastructure specifics. This abstraction simplifies storage management and allows for better resource utilization and control while supporting various storage backends, including cloud provider-specific storage systems and on-premises solutions.

Code:

```yaml
storageclass-eks.yaml
apiVersion: storage.k8s.io/v1
kind: StorageClass
metadata:
  name: redis-storage-eks
provisioner: kubernetes.io/aws-ebs
parameters:
  type: gp3
```

StatefulSet and PVC

A Kubernetes StatefulSet utilizing a PVC. Notice the **volumeClaimTemplates** section, this dynamically creates the **PersistenVolumeClaim** and targets the **StorageClass** we previously defined.

Code:

```yaml
Statefulset.yaml
apiVersion: apps/v1
kind: StatefulSet
metadata:
```

```yaml
  name: redis
  namespace: redis
spec:
  replicas: 1
  selector:
    matchLabels:
      app: redis
  serviceName: redis
  template:
    metadata:
      labels:
        app: redis
    spec:
      configMap:
        name: redis-config
      containers:
      - env:
        - name: REDIS_CONF_FILE
          value: /conf/redis.conf
        image: redis:latest
        name: redis
        ports:
        - containerPort: 6379
          name: redis
        volumeMounts:
        - mountPath: /conf
          name: redis-config
        - mountPath: /data
```

```
          name: redis-data
     volumes:
     - name: redis-config
volumeClaimTemplates:
- metadata: null
  name: redis-data
  spec:
    accessModes:
    - ReadWriteOnce
    resources:
      requests:
        storage: 1Gi
      storageClassName: redis-storage-eks
```

Create the resources

When deploying a Redis instance on Kubernetes, it is essential to create the resources in the following order to ensure proper configuration and avoid dependency issues:

- **Namespace:** Create the namespace to logically isolate and manage resources related to Redis:

    ⟩ kubectl apply -f namespace.yaml

    *Note: Set your context to your new namespace: "kubectl config set-context --current --namespace=redis"

- **ConfigMap**: Create the **ConfigMap** containing the Redis configuration settings. The **ConfigMap** should be created before the StatefulSet, as the StatefulSet references it for its configuration:

    ⟩kubectl apply -f configmap.yaml

- **StorageClass:** Create the **StorageClass** for the respective cloud provider (EKS or GKE). This should be done before creating the StatefullSet, as it will create the PVC which references the **StorageClass** for dynamic provisioning:

    ⟩kubectl apply -f storageclass-eks.yaml

- **StatefulSet:** Create the **StatefulSet** that deploys the Redis instances. The **StatefulSet** must be created after the **ConfigMap**, **StorageClass**, as it references all of these resources:

    ⟩kubectl apply -f statefulset.yaml

Verify the Redis deployment

After waiting a few minutes for all the resources to be created by the cloud provider, we can check that they are properly created and in the correct status.

Code:

```
>kubectl get configmaps,storageclass,statefulset,pod
```

NAME	DATA	AGE
configmap/kube-root-ca.crt	1	22h
configmap/redis-config	1	21h

NAME	PROVISIONER
storageclass.storage.k8s.io/gp2 (default)	kubernetes.io/aws-ebs
storageclass.storage.k8s.io/redis-storage-eks	kubernetes.io/aws-ebs

NAME	READY	AGE
statefulset.apps/redis	1/1	20h

NAME	READY	STATUS	RESTARTS	AGE
pod/redis-0	1/1	Running	0	20h

Test the Redis deployment

From your local machine, you can set up a port forward which will create a bridge between the Redis pod and you. You set this up in terminal, as follows:

```
>kubectl port-forward pod/redis-0 6379:6379
Forwarding from 127.0.0.1:6379 -> 6379
Forwarding from [::1]:6379 -> 6379
```

Now you can make a call directly to the pod in another terminal window using Netcat:

```
>echo PING | nc localhost 6379
+PONG
```

```
>echo INFO | nc localhost 6379
# Server
redis_version:7.0.11
...
> echo "GET foo " | nc localhost 6379
$-1
> echo "SET foo bar" | nc localhost 6379
+OK
> echo "GET foo " | nc localhost 6379
Bar
```

Expose the Redis database

So now we have a functional and tested Redis deployment in our EKS cluster. We now want to expose the service so that it is available to services outside of the cluster. This is a crucial next step to setting up a multi-cloud deployment of Redis!

To expose your Redis deployment running in EKS to services outside of the cluster, you can create a Service with type **loadBalancer**. This will expose the Redis instance through a cloud provider's load balancer and assign a public IP address or DNS name to it.

Create a YAML file for the **loadBalancer** Service, as follows:

apiVersion: v1

kind: Service

metadata:

  name: redis-lb

  namespace: redis

redis-service-lb.yaml: null

spec:

  ports:

  - port: 6379

  selector:

```
    app: redis
  type: LoadBalancer
```

Apply the Service YAML to your EKS cluster:

```
>kubectl apply -f redis-service-lb.yaml
```

service/redis-lbcreated

Get the public IP address or DNS name of the **loadBalancer** created by the cloud provider for the Service:

```
>kubectl get svc -n redis --output=jsonpath='{.status.loadBalancer.ingress[].hostname}'
```

"afdc29e56e2f84725a2432bcfc35e148-2117135052.us-east-2.elb.amazonaws.com"

You will see the public IP address or DNS name of the **loadBalancer**.

Now, the Redis service is exposed and accessible from outside the cluster. You can use the public IP address or DNS name to connect to it from services running outside the EKS cluster.

Keep in mind that exposing Redis over the public internet might pose security risks. In further chapters, we will mitigate these risks in the following ways:

- Use encryption for data-in-transit by enabling SSL/TLS for the Redis instances. You will need to configure the Redis instances to use SSL/TLS and provide the necessary certificates and keys.
- Use authentication to protect access to the Redis instances. Set up a password for the Redis instances and configure the clients to use the password when connecting to the instances.
- Restrict access to the **loadBalancer** by using security groups, firewall rules, or any other cloud-provider-specific access control mechanisms to only allow traffic from the IP addresses of the nodes or the services that require access to the Redis instance.

Test out the public endpoint

We can test out the public endpoint in much the same way. Make sure the port-forward has been terminated (Control-c), then us Netcat in the same way as before but using the public endpoint:

```
> echo "SET K8s Rocks! " | nc afdc29e56e2f84725a2432bcfc35e148-2117135052.us-east-2.elb.amazonaws.com 6379
```

+OK

```
> echo "GET K8s " | nc afdc29e56e2f84725a2432bcfc35e148-2117135052.us-east-2.elb.amazonaws.com 6379
```

Rocks!

Conclusion

In this chapter, we have learned how to deploy a highly available Redis instance. We created and deployed Kubernetes resources such as **Namespace**, **ConfigMap**, **StorageClass**, **PersistentVolume**, **PersistentVolumeClaim**, and **StatefulSet** to set up the Redis instance in EKS. We then exposed the Redis service to the outside world using a **LoadBalancer** Service so it can be accessed by external services and other Redis instances running in different cloud environments like GKE. This foundational understanding of deploying stateful applications in Kubernetes sets the stage for more complex deployments.

Congratulations on your progress! You have taken significant strides in understanding and setting up a multi-cloud Kubernetes environment.

In the next chapter, we will shift our focus to Stateless application-based clusters. We will continue building on the multi-cloud Redis exercise, adding more elements to enhance our understanding of Kubernetes and its capabilities in handling various types of applications. We will explore the differences between stateless and stateful applications, learn how to deploy, scale, and manage stateless applications using Kubernetes resources like Deployments and ReplicaSets. By the end of the next chapter, you will have a solid understanding of how to deploy and manage both stateful and stateless applications in a multi-cloud environment, further strengthening your skills in Kubernetes and cloud-native technologies.

Join our book's Discord space

Join the book's Discord Workspace for Latest updates, Offers, Tech happenings around the world, New Release and Sessions with the Authors:

https://discord.bpbonline.com

CHAPTER 5
Stateless Application Kubernetes Cluster Design

Introduction

In this chapter, we will delve into the nuances of architecting, designing, and implementing Kubernetes clusters tailored for stateless applications. Stateless applications, unlike their stateful counterparts, do not store or maintain any persistent data, making their design patterns unique and essential for specific use cases. This chapter will provide valuable insights into the design principles, best practices, and deployment strategies to effectively run stateless applications in a multi-cloud Kubernetes environment.

Structure

Here is an overview of the topics we will cover:
- Overview of stateless applications
- Architecture and design considerations
- Kubernetes resources for stateless applications
- Load balancing and scaling stateless applications
- Stateless application deployment patterns
- Hands-on system design and project plan
- Project plan for deploying NGINX web server on EKS

Objectives

In this chapter, our objectives are to provide readers with a comprehensive understanding of stateless application Kubernetes cluster design in a multi-cloud environment. By the end of this chapter, readers should be able to grasp the fundamental concepts of stateless applications, recognize the unique challenges and opportunities presented by multi-cloud deployments, and apply best practices for designing, implementing, and managing stateless application clusters across different cloud providers. Furthermore, we aim to equip readers with practical knowledge of various tools, techniques, and Kubernetes resources specific to stateless applications, enabling them to make informed decisions when architecting solutions for their own projects.

Overview of stateless applications

In this section, we will provide an overview of stateless applications running in multi-cloud Kubernetes environments. Stateless applications are those that do not store any client-specific data on the server between requests, which makes them inherently more scalable and easier to manage in distributed environments like multi-cloud Kubernetes.

Key characteristics of stateless applications are as follows:
- No client-specific data stored on the server between requests.
- Easy horizontal scaling due to the lack of state management requirements.
- Simplified deployment and management in distributed environments.
- Independence of requests, meaning they can be handled by any available instance.
- Enhanced fault tolerance because of application instances being interchangeable.

Some common examples of stateless applications are:
- **User management service:**
  - **Responsibilities**: Handles user registration, authentication, and profile management.
  - **External storage**: User data is stored in a centralized database or a cloud-based authentication service (for example, AWS Cognito, Auth0), ensuring centralized management and accessibility.
  - **Stateless design**: The service retrieves and processes user data from the external storage as needed, without storing it locally.
- **Order management service:**
  - **Responsibilities**: Processes customer orders, manages inventory, and coordinates fulfillment.

- o **External storage**: Order data is stored in a persistent database like MySQL or PostgreSQL, or a distributed database like Cassandra or MongoDB, providing data durability and consistency.
- o **Stateless design**: The service interacts with the external storage to create, update, and manage orders, but doesn't hold any order data locally.
- **Product catalog service:**
  - o **Responsibilities**: Manages product information, search functionality, and pricing.
  - o **External storage**: Product information is stored in a database or a cache like Redis or Memcached, enabling efficient retrieval and updates.
  - o **Stateless design**: The service retrieves product data from the external storage upon request, leveraging caching for frequently accessed information, but avoiding local storage.

The advantages of stateless applications in multi-cloud Kubernetes are as follows:
- Easier scalability due to the lack of state management requirements.
- Improved fault tolerance as application instances can be replaced without affecting overall functionality.
- Simplified deployments and updates as there are no dependencies on previous states.
- Better resource utilization as each instance can handle any incoming request.
- Simplified data replication and synchronization across multiple cloud providers.

The challenges of stateless applications in multi-cloud Kubernetes are:
- Ensuring data consistency across multiple cloud providers in real-time.
- Managing latency and network-related issues when communicating between cloud providers.
- Designing and implementing a proper load balancing and traffic management strategy.
- Ensuring data security and compliance when dealing with sensitive data across different cloud providers.
- Addressing potential vendor lock-in concerns and ensuring seamless integration of services across multiple cloud providers.

Architecture and design considerations

In this section, we will discuss the various components and considerations that come into play when designing stateless application Kubernetes clusters. These components can be broadly categorized into four main areas: application design, Kubernetes resources, networking, and storage. Let us know more about application design in the next section.

Designing for scalability

Stateless applications should be designed to take full advantage of Kubernetes' capabilities for horizontal scaling. The application should be able to handle any number of instances running concurrently and processing requests independently. This allows Kubernetes to scale the application up and down efficiently according to demand and resource utilization.

Data persistence and management

In a stateless application, data is not stored locally within the application or its containers. Instead, any necessary data should be stored in a separate, persistent storage service. This could be a database or a cloud-based storage service. In a multi-cloud context, it is important to consider data management strategies that allow for efficient access to this data across different cloud environments. This might involve using multi-region databases or implementing a data caching layer.

Load balancing and networking

In a multi-cloud environment, effective load balancing becomes crucial. The application should be designed to work well with Kubernetes services, which automatically distribute traffic among pods. Due to the stateless nature of the workload the number of replicas running concurrently should not affect the application functionality.

Fault tolerance and health checks

Fault tolerance is a key aspect of stateless application design. The application should be designed to handle failures gracefully, and to recover quickly when they occur. This involves implementing robust health checks to allow Kubernetes to monitor the health of the application and restart or reschedule pods as necessary.

Containerization and image design

Container images for stateless applications should be designed to be lightweight, efficient, and secure. The application code, runtime, system tools, system libraries, and settings should all be packaged into the container image. In a multi-cloud environment, it is particularly important to ensure that container images are portable across different cloud platforms. Typically vendor published images have multiple architecture support, are cloud agnostic and are as discreet as possible. Some examples are NGINX (**https://hub.docker.com/_/nginx**) or Redis (**https://hub.docker.com/_/redis**).

Configuration management

Stateless applications should be designed to consume configuration data in a way that is decoupled from the application code. This can be achieved with environment variables at both build and run time. In a multi-cloud context, consider using a centralized configuration service to manage configuration data across different cloud environments. A runtime example is Hashicorp Consul where an agent connects to a centrally managed configuration store and sets environment variables. Most CI/CD tools like GitHub Actions, GitLab, Jenkins, and so on, can inject variables into container images during the build process.

Interoperability and portability

In a multi-cloud environment, interoperability and portability become key concerns. Stateless applications should be designed to be cloud-agnostic as much as possible, and to work well with Kubernetes' abstraction of underlying cloud resources. This will make it easier to deploy and manage the application across different cloud platforms. For example, taking our User Management service, we would include libraries in the application which can support all cloud related services, not just one specific vendor.

Kubernetes resources

Let us know more about Kubernetes resources:

- **Scalability and load balancing**: One of the fundamental characteristics of stateless applications is that they can scale horizontally in response to changes in load. In a Kubernetes cluster, this is achieved by increasing or decreasing the number of pods running the application. Load balancing, handled by services in Kubernetes, distributes traffic evenly among these pods, ensuring that no single pod becomes a bottleneck.
- **Fault tolerance**: Fault tolerance is another key aspect to consider in the design of stateless applications on Kubernetes. By running multiple replicas of an application and distributing them across different nodes, Kubernetes ensures that an application remains available even if a pod or node fails. With proper health checks and liveness probes in place, Kubernetes can automatically restart or reschedule pods as needed, minimizing downtime.
- **Deployments and ReplicaSets**: For stateless applications, Deployments and ReplicaSets are key Kubernetes resources. A deployment manages a ReplicaSet, which in turn ensures that a specified number of pod replicas are running at any

given time. Deployments manage the rollout and rollback of changes to your application, ensuring zero downtime and maintaining the availability of your application during updates.

- **Horizontal pod autoscalers**: For applications that experience variable load, **Horizontal Pod Autoscalers (HPA)** are a vital component. HPAs adjust the number of pod replicas in a ReplicaSet or Deployment based on observed CPU utilization, or on custom metrics supported by the Metrics Server. This allows your application to respond to changes in load dynamically.
- **Container images**: The design of the container image is also a critical aspect in stateless applications. The container image should be designed to be stateless and immutable, meaning it does not change once it is running. Any necessary state should be stored outside of the container, for example in a database or a remote storage system.
- **Environment variables and ConfigMaps**: Configuration data for stateless applications is typically provided through environment variables, which can be set individually for each pod, or through ConfigMaps, which allow you to decouple configuration artifacts from image content. This makes your applications more portable and easier to manage.
- **Multi-cloud considerations**: In a multi-cloud environment, additional considerations may come into play. These can include dealing with different networking and storage configurations across providers, as well as managing the distribution of application components and data across regions or cloud providers for reasons such as data sovereignty, latency, or cost.

Network

Let us know more about network:

- **Service discovery and DNS**: Stateless applications in Kubernetes rely heavily on the concept of services for service discovery. Services provide a stable network endpoint for pods in the cluster. In multi-cloud environments, cross-cluster service discovery may be required, and can be achieved through various methods including federated services or multi-cluster services. DNS configuration plays a crucial role in this, as it allows services to be discovered using their service names instead of relying on IP addresses, which can change frequently in a dynamic environment like Kubernetes.
- **Load balancing**: Load balancing is essential for distributing network traffic across multiple pods, ensuring high availability and reliability. In a multi-cloud environment, you may need to consider the different load balancing solutions provided by different cloud providers. Kubernetes provides its own load balancer type called **Service**, and in a multi-cloud setup, you can use a combination of Kubernetes Service objects and cloud-provider-specific load balancers.

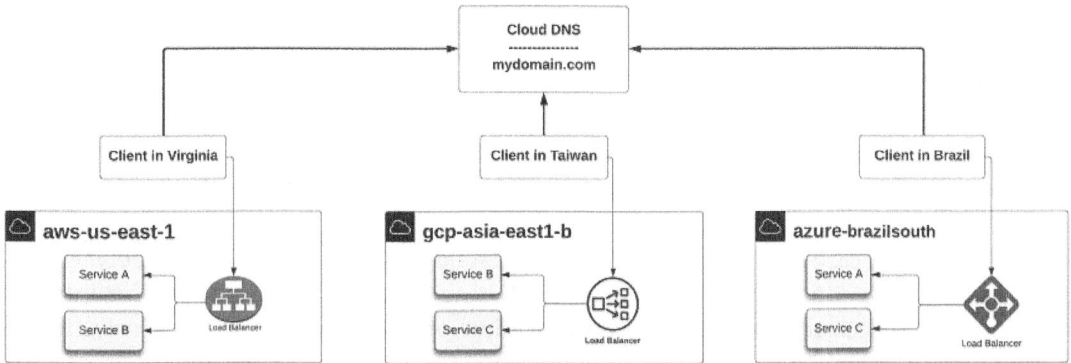

Figure 5.1: Global DNS and Load Balancing

- **Ingress and egress controls**: Ingress and egress controls determine how incoming and outgoing traffic is managed. In a multi-cloud setup, designing these controls involves understanding and managing the networking rules of each cloud provider. Kubernetes provides Ingress objects for managing incoming HTTP(s) traffic. For egress, Network policies can be used to control outgoing traffic at the pod level.

- **Network policy and security**: Network policies in Kubernetes allow you to define how groups of pods are allowed to communicate with each other and other network endpoints. In a multi-cloud environment, you will need to ensure these policies are consistently applied across your clusters, regardless of the underlying cloud provider. This can be a complex task, as different cloud providers may have different default network policies. In later chapters, we will visit more complex ways of managing network and security policies via a Service Mesh

- **Network performance and optimization**: Network performance can greatly affect the performance of stateless applications, especially when dealing with high traffic volumes. In a multi-cloud environment, consider the network performance characteristics (latency, throughput, and so on) of each cloud provider and how this might affect application performance. Various optimization techniques such as fine-tuning TCP parameters, utilizing CDN services, or using network plugins for Kubernetes can help to enhance network performance.

- **Inter-cluster and cross-region networking**: In a multi-cloud environment, your Kubernetes clusters are likely spread across multiple regions or even different cloud providers. Designing the network architecture for such a setup involves setting up secure and reliable network connectivity between these clusters. This might involve using technologies such as VPNs, dedicated interconnects, or even Kubernetes-native solutions like Submariner.

Remember, network design for stateless applications in a multi-cloud environment can be complex and requires careful planning and consideration of the unique characteristics and capabilities of each cloud provider.

Kubernetes resources for stateless applications

Kubernetes offers a suite of resources designed to facilitate the orchestration of both stateful and stateless applications. However, the ephemeral and interchangeable nature of stateless applications calls for a slightly different approach when it comes to resource utilization. In the case of stateless applications, data is not preserved beyond the life of the pod, and any pod can respond to a request indistinguishably from another. This intrinsic behavior influences the choice and configuration of Kubernetes resources used in the deployment of stateless applications.

Kubernetes resources that are particularly relevant for managing stateless applications include the following:

- **Pods**: These are the smallest deployable units of computing that can be created and managed in Kubernetes. Pods are ephemeral and can be replaced at any moment, making them an excellent match for stateless applications. Since stateless applications only rely on external storage pods are typically lightweight without locally attached storage.
- **Deployments**: A Deployment provides declarative updates for Pods and ReplicaSets. You describe a desired state in a Deployment, and the Deployment Controller changes the actual state to the desired state at a controlled rate.
- **ReplicaSets**: A ReplicaSet's purpose is to maintain a stable set of replica Pods running at any given time. As such, it is often used to guarantee the availability of a specified number of identical Pods.
- **Services**: A Kubernetes Service is an abstract way to expose an application running on a set of Pods as a network service. For stateless applications, services provide a stable network endpoint, which is crucial given that pods are ephemeral and can be replaced at any moment.
- **Ingress**: An API object that manages external access to the services in a cluster, typically HTTP. Ingress may provide load balancing, SSL termination and name-based virtual hosting. This is particularly important for stateless applications which often need to handle large volumes of incoming traffic.
- **Vertical pod scaling**: When vertically scaling pods we focus on the available resources allocated to the application. For example, take a NGNX web server that is running on 100 millicores and 512 MB of memory which is experiencing slow request processing times, or even more dramatic repeated failures due to OOM errors. You can vertically scale the pod by allocating more resources, say doubling the millicore count to 250 and the memory to 1 GB.

- **HPA**: HPA automatically scales the number of pods in a replication controller, deployment, replica set or stateful set based on observed CPU utilization. This is particularly useful for stateless applications that might need to scale up and down quickly based on demand. Keep in mind this process depends on additional applications to be deployed to make it work, a popular example is KEDA (**https://keda.sh/**).
- **Namespaces**: Namespaces are intended for use in environments with many users spread across multiple teams, or projects. They provide a scope for names and can be helpful in dividing cluster resources between multiple uses.

By leveraging these resources, Kubernetes can efficiently orchestrate stateless applications, even in a complex, multi-cloud environment. In the following sections, we will delve deeper into the specific usage and benefits of these resources for stateless applications.

Load balancing and scaling stateless applications

Stateless applications, by their very nature, are primed for horizontal scaling — that is, increasing the number of instances to accommodate increased load. This characteristic is particularly advantageous in multi-cloud environments where workloads can be distributed across multiple cloud platforms for better resource utilization, cost-effectiveness, and even geographical proximity to users.

Load balancing

In a multi-cloud Kubernetes environment, load balancing becomes a critical factor. Load balancing ensures that incoming network traffic is efficiently distributed across multiple pods in a cluster, optimizing resource utilization, maximizing throughput, reducing latency, and ensuring fault tolerance. Kubernetes provides a built-in service type known as LoadBalancer, which can leverage cloud provider's load balancers to distribute external traffic to the pods within a cluster. In a multi-cloud scenario, it is also possible to use global load balancers, which can distribute traffic across multiple clusters in different regions or cloud providers.

Pod scaling

Pod scaling or vertical scaling is the process of adjusting the number of active pods to match the demand for your application. In Kubernetes, ReplicaSets and Deployments are the main resources for managing pod scaling. You define the desired number of pods, and Kubernetes ensures that this number of pods is always running, replacing any that fail or terminate.

Horizontal Pod Autoscaling

Horizontal Pod Autoscaling is a mechanism that automatically adjusts the number of pods in a replication controller, deployment, or replica set based on observed CPU utilization (or, with custom metrics support, on some other application-provided metrics). This is particularly important for stateless applications, which can be scaled horizontally in response to changing demand.

Cluster Autoscaling

While HPA adjusts the number of pods to match the load, Cluster Autoscaler adjusts the size of the cluster itself to fit the current needs of your applications. When a new pod cannot be scheduled due to lack of resources, Cluster Autoscaler adds a new node to the cluster. Conversely, if some nodes are underutilized and all pods could be scheduled even without them, Cluster Autoscaler removes the underutilized nodes from the cluster. When running in a multi-cloud environment, Cluster Autoscaler should be properly configured for each cloud provider to function optimally.

Auto healing

Given the ephemeral nature of stateless applications, they need to be designed for failure. Kubernetes, by design, supports self-healing capabilities. For instance, when a pod fails, the Kubernetes scheduler ensures that a new pod is spun up either in the same node or in a different node, ensuring the application's availability. Kubernetes relies on the liveness and readiness probe to detect the state of the application in the pod. The checks are the responsibility of the applications and will be unique to the application. For example, a NGINX web application might have an endpoint such as `/health`, or `/ready` exposed on port `80` while a Flask Python application may have a `/live` endpoint exposed on port `3000`. The application must define this as part of their configuration.

In summary, properly designed and managed, stateless applications in a multi-cloud Kubernetes environment can provide high availability and efficient resource utilization. However, effective load balancing and auto-scaling require careful planning and an understanding of both the application requirements and the features and constraints of the underlying Kubernetes platform and cloud infrastructure.

Stateless application deployment patterns

Designing and deploying stateless applications in Kubernetes can follow several patterns, each with its own advantages and implications. The choice of pattern often depends on the specific requirements of your application and your operational environment. Here, we will discuss two primary patterns: deploying to a single cloud and deploying to a multi-cloud environment.

Single cloud deployment pattern

In this pattern, your stateless application is deployed to a single Kubernetes cluster that resides within one cloud provider's environment. This is often the starting point for most organizations. It is simpler and allows you to leverage the specific tooling, services, and advantages of a single cloud provider. However, it also means that your application is tied to the availability, performance, and cost structure of that specific cloud provider.

Blue-green deployment

This is a strategy for releasing applications by shifting traffic between two identical environments running different versions of the application. Using a mechanism such as DNS or Gateway routing to direct all traffic to the blue environment while deploying to the green environment allows for zero downtime as well as smoke tests on the newly deployed code in a real production environment without affecting end users. When the green deployment is deemed ready to go live traffic can be shifted from the blue to the green environment. After switching traffic and everything is stable, the blue environment can be decommissioned, or kept on standby for rollback if needed.

Canary deployment

Canary deployments, named after the miners' practice of using canaries to detect poisonous gases, are a software deployment strategy that uses a controlled rollout of a new version to a subset of users before releasing it to everyone. It's like dipping your toes in the water before diving headfirst. Deployment is done to a new set of instances and a small percentage of user traffic is routed to it. While monitoring the traffic to the new code, if everything is working as intended the amount of traffic is gradually increased until it serves all users. If any issues are found the traffic can be reduced back to zero and the issues can be fixed before trying again.

Multi-cloud deployment pattern

As your application and organization mature, you may choose to deploy your stateless application across multiple Kubernetes clusters that reside in different cloud environments. This pattern is more complex but provides several advantages, including avoiding vendor lock-in, increasing availability and disaster recovery, and the potential to optimize costs by using resources from different providers. The following figure depicts multi-cloud deployment:

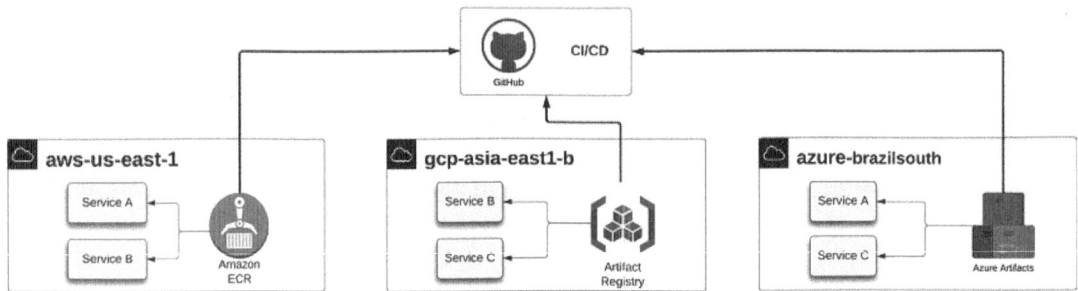

Figure 5.2: Multi-cloud Deployment

These deployment patterns can be used with the load balancing and autoscaling strategies discussed earlier to design a resilient, performant, and cost-effective stateless application deployment on Kubernetes, whether you are using a single cloud or a multi-cloud environment.

Hands-on system design and project plan

One of the most common examples of stateless applications is a web server, such as Nginx or Apache HTTP Server. For the purpose of our hands-on system design and project plan, we will use Nginx due to its popularity, performance, and ease of use. Running a web server like Nginx in a containerized, multi-cloud environment is a typical use case and allows us to explore various aspects of deploying and managing stateless applications across multiple Kubernetes clusters.

We will be following the given outline for our hands on exercise:

- **Introduction and example application overview**
    - Brief introduction to the example application (Nginx web server)
    - Challenges of deploying NGINX in a multi-cloud Kubernetes environment
    - Why this application is representative of common stateless applications
- **System design for stateless application**
    - Architecture overview
        - High-level architecture of Nginx deployment in multi-cloud environment
        - Role of Kubernetes in managing the deployment
    - Networking design
        - Designing network for optimal performance and availability
        - Handling network traffic between different cloud providers
    - Load balancing and scaling
        - Strategies for load balancing

- Scaling considerations for stateless applications
- **Project plan for stateless application deployment**
    - Project scope, objectives, and key deliverables
        - Defining the purpose and expected outcomes of the project
    - Stakeholders' roles and responsibilities
        - Identifying key project members and their duties
    - Project timeline and milestones
        - Key stages of the project (planning, design, implementation, testing, deployment)
        - Important dates and milestones
    - Resource allocation
        - Personnel, hardware, software requirements for each stage
    - Risk management plan
        - Identification of potential risks and their mitigation strategies
- Implementation steps
    - Step-by-step guide for deploying the Nginx server on multi-cloud Kubernetes clusters
        - Setting up necessary Kubernetes resources (namespaces, services, deployments, etc.)
        - Network configurations
    - Testing and validation steps
    - Monitoring and maintenance considerations
- Conclusion
    - Recap of the key takeaways from the hands-on example
    - How the learned principles can be applied to other stateless applications
    - Final thoughts and next steps

Brief introduction to the example application

The example application that we will be discussing in this chapter is a NGINX web server. NGINX is a high-performance, open-source software for web serving, reverse proxying, caching, load balancing, media streaming, and more. It is known for its high performance, stability, rich feature set, simple configuration, and low resource consumption.

In the context of Kubernetes, NGINX often serves as an Ingress controller, directing HTTP and HTTPS traffic to different services within the cluster based on routing rules. However, it can also serve as a stand-alone web server within a Pod, serving static content or acting as a reverse proxy. This is how we will be deploying NGINX for our exercise.

In a multi-cloud environment, deploying a NGINX web server involves creating Docker containers to run NGINX, setting up Kubernetes deployments to manage these containers across clusters in different cloud environments, and configuring services and ingress resources to expose the NGINX servers to the internet.

Through this hands-on example, we aim to provide a practical understanding of how to design, deploy, and manage stateless applications in a multi-cloud Kubernetes environment.

Challenges in deploying applications

Deploying applications in a multi-cloud Kubernetes environment involves several challenges such as the following:

- **Configuration consistency**: Ensuring consistent configuration across multiple cloud environments can be complex. Each cloud provider has its own set of tools, services, and interfaces, which can lead to discrepancies and potential errors.
- **Networking**: Setting up networking that spans across multiple cloud providers is a challenge due to variations in network implementation across different cloud platforms. This includes setting up load balancers, ingress controllers, and service discovery.
- **Security**: Multi-cloud deployments expand the security perimeter, which can increase the risk of attacks. Ensuring secure communication and data protection across clouds requires careful planning and implementation of security policies and controls.
- **Monitoring and troubleshooting**: Monitoring applications across different clouds can be complex due to different logging and monitoring tools provided by each cloud vendor. Aggregating and analyzing logs and metrics from different sources is a challenge.
- **Cost management**: Managing costs can become complicated due to the different pricing models used by each cloud provider. It requires careful monitoring and management to avoid unexpected charges.
- **Data sovereignty and compliance**: Data stored in different geographical locations can be subject to different regulations, which can complicate compliance management.
- **Auto-scaling**: While Kubernetes natively supports auto-scaling, configuring it optimally for a NGINX server across multiple clouds to maintain performance and minimize costs can be challenging.
- **Skills and knowledge**: The team needs to have expertise in Kubernetes as well as each of the cloud platforms being used. This multi-domain knowledge can be a challenge to acquire and maintain.

Why this application is representative of common stateless applications

The NGINX web server is representative of common stateless applications for a number of reasons. Firstly, it is designed to handle each request independently of others, making it inherently stateless. This means that no session information is stored by the server between requests. Instead, every request is processed based solely on the information that comes with it, which aligns perfectly with the stateless nature of HTTP protocol that web servers typically use.

Secondly, it is highly scalable and resilient. As a stateless application, new instances of a NGINX server can be created and destroyed on-demand without affecting the overall operation of the application. This aligns well with Kubernetes' core principles of scaling and self-healing.

Finally, NGINX is widely used in a variety of settings, making it a familiar and practical example for many developers and system administrators. Its functionality can range from serving static websites to acting as a reverse proxy, making it a versatile choice for demonstrating deployment strategies in a multi-cloud Kubernetes environment.

Thus, understanding the deployment and management of a NGINX server in a multi-cloud Kubernetes environment can provide valuable insights applicable to a wide range of stateless applications.

Multi-cloud NGINX web server system design

In this section, we will dive into the system design for our NGINX web server deployment in a multi-cloud Kubernetes environment, focusing on four main components: Architecture Overview, Network Design, Load balancing and Scaling. We will consider the differences between EKS and GKE for each of these components.

Architecture overview

The architecture of a typical deployment of a NGINX web server in a multi-cloud Kubernetes environment revolves around several key components. First, we deploy the NGINX web server as a pod within our Kubernetes cluster. This pod encapsulates our NGINX application and provides a layer of abstraction that allows it to run consistently across different cloud environments. The pod is deployed within a specific namespace, which provides a scope for managing resources associated with the NGINX server. To expose the NGINX server to network traffic, we use a Kubernetes service. This service acts as a stable endpoint for the pod, even as the pod itself may be replaced or moved due to scaling or failure recovery operations. To manage external traffic coming into our cluster, we deploy an Ingress controller. The Ingress controller routes external requests to the appropriate services based on rules defined in an Ingress resource. While the basic architecture remains the same, different cloud providers may offer additional features or integrations. For example, AWS's EKS and Google's GKE offer integration with their respective load balancer and network services, which can be utilized for better performance and ease of management.

Network design

The network design of deploying a NGINX web server in a multi-cloud Kubernetes environment is a critical aspect of the architecture. The goal is to ensure optimal performance, security, and reliability. Each cloud provider (like AWS's EKS or Google's GKE) comes with its own set of networking features, but Kubernetes provides a consistent networking model across different environments. All pods in a Kubernetes cluster are assigned their own IP address and can communicate with each other without the need for NAT. Services provide a stable IP address and DNS name that other pods can use to access a given application, like our NGINX web server. In a multi-cloud environment, it is essential to manage ingress and egress traffic carefully. This is where Ingress controllers and Network Policies come into play. An Ingress controller, such as NGINX or Traefik, is responsible for managing incoming traffic and routing it to the appropriate services based on rules defined in an Ingress resource. Network Policies, on the other hand, allow us to define rules for how pods communicate with each other and other network endpoints, providing an additional layer of security. In a multi-cloud setup, it is also crucial to consider inter-cluster communication and network latency between clusters in different cloud environments. Tools like service meshes (for example, Istio, Linkerd) or dedicated interconnects provided by the cloud providers can help manage these challenges.

Load balancing

Load balancing in a multi-cloud Kubernetes environment for a stateless application like a NGINX web server is a critical part of ensuring high availability and reliable performance. In Kubernetes, the Service resource is the primary means of load balancing traffic to Pods. When a service is defined, Kubernetes automatically handles the distribution of network traffic between Pods that are part of the service. This ensures that incoming requests to the NGINX web server are efficiently distributed across all available instances.

In a multi-cloud scenario, however, additional considerations come into play. Each cloud provider may offer its own load balancer solution, such as AWS **Elastic Load Balancer (ELB)** or Google Cloud Load Balancer. These cloud provider-specific solutions can be automatically provisioned by Kubernetes as a Service of type LoadBalancer, which will handle routing external traffic to the appropriate Pods in the cluster.

Furthermore, a global load balancer may be required to distribute traffic between clusters in different cloud environments. This is particularly relevant if you are implementing a multi-region or multi-cloud strategy for redundancy and high availability. In this case, DNS-based global traffic management solutions can be used to route traffic to the closest or best-performing Kubernetes cluster serving the NGINX web servers.

Despite the stateless nature of NGINX web servers, session affinity (or "sticky sessions") may still be a requirement for certain use cases. Kubernetes Services support session affinity based on ClientIP, ensuring that all requests from a particular client are sent to the same Pod, as long as that Pod remains running.

These strategies combined ensure a reliable and resilient setup for load balancing a NGINX web server deployment across multiple cloud environments.

Scaling

Scaling is a vital factor when deploying a NGINX web server in a multi-cloud Kubernetes environment. The stateless nature of a NGINX web server allows for easy horizontal scaling, meaning that new instances of the application can be added or removed dynamically based on demand. In Kubernetes, this scaling process can be automated using the HPA, which adjusts the number of running instances (pods) of an application based on observed CPU utilization or other selected performance metrics.

For multi-cloud environments, Kubernetes Federation can be used to manage scaling across different clusters in different regions or cloud providers. The federation control plane can replicate resources across member clusters and make global adjustments to the number of replicas as needed. This way, it can ensure that the application scales up during high demand and scales down when demand is low, optimizing resource usage and cost.

Furthermore, Kubernetes also supports Cluster Autoscaling. This feature automatically adjusts the size of a Kubernetes Cluster so that all pods have a place to run and there are no unneeded nodes. This is crucial when dealing with sudden bursts of traffic or varying load patterns.

Finally, the Kubernetes **Vertical Pod Autoscaler (VPA)** can be used to automatically adjust the CPU and memory reservations for your NGINX pods, ensuring optimal resource allocation and performance.

It is important to remember that scaling strategies should be designed with the specifics of each cloud provider in mind, as each provider may have different limits or best practices for scaling. Overall, a well-implemented scaling strategy ensures that your NGINX web server deployment can handle varying loads efficiently across multiple clouds.

Project plan for deploying NGINX web server on EKS

Let us go over the project plan now.

Project plan

Here is a great project plan outline you can use for deploying NGINX web server as a stateless application on both EKS and GKE. Note that every company is different so use these as guidelines and fill in the appropriate content for your business use case.

Project scope

The scope of this project is to deploy a NGINX web server as a stateless application on existing Kubernetes clusters in both EKS and GKE environments. The deployment will be designed to handle dynamic scaling and load balancing to ensure high availability and performance.

Objectives

The objectives are as follows:
- Design and implement a robust NGINX web server deployment architecture for both EKS and GKE.
- Deploy the NGINX web server on EKS and GKE
- Test the implementation to ensure it meets performance and availability criteria.

Key deliverables

The key deliverables are as follows:
- Architecture design document for NGINX deployment on EKS and GKE
- Kubernetes manifests for NGINX deployment on EKS and GKE.
- Network connectivity configuration and documentation.
- Test plans and test results.
- Deployment and maintenance guides.

Project stakeholders and roles

The project stakeholders and roles are as follows:
- **Project manager:** Oversees the project, ensures milestones are met, and manages resources.
- **Cloud architect:** Designs the architecture and ensures best practices for multi-cloud deployment.
- **Kubernetes engineer:** Implements and deploys the NGINX instances on EKS and GKE.
- **Network engineer:** Establishes and maintains network connectivity between EKS and GKE.

- **QA engineer:** Develops and executes test plans to validate the deployment.

Project timeline

The project timeline is as follows:
- **Planning (1 week):** Define project objectives, scope, and deliverables. Identify stakeholders and their roles.
- **Design (2 weeks):** Design the architecture and networking for NGINX deployment on EKS and GKE.
- **Implementation (3 weeks):** Develop Kubernetes manifests, configure storage, and establish network connectivity.
- **Testing (2 weeks):** Execute test plans, validate functionality, and address any issues found.
- **Deployment (1 week):** Deploy the final NGINX instances on EKS and GKE and provide documentation.

Resources

The resources are as follows:
- **Personnel:**
  - **Project manager (PM):** Oversees the entire project, ensuring deliverables are met on time, within budget, and to scope. Manages the team, stakeholders, and communication.
  - **Cloud architect:** Designs and implements the cloud infrastructure, considering scalability, security, and cost-effectiveness. Chooses cloud providers and services.
  - **Kubernetes engineer:** Deploys, manages, and optimizes containerized applications using Kubernetes. Ensures smooth operation and high availability.
  - **Network engineer:** Designs, configures, and maintains the network infrastructure, ensuring connectivity, security, and performance.
  - **QA engineer:** Tests the application for functionality, usability, and performance. Identifies and reports bugs before release.
- **Software:** Kubernetes, NGINX, KubeCTL, AWS CLI, Google Cloud SDK

Risk management plan

Let us now go over the various risks that may arise and their mitigation:
- **Risk 1:** Unexpected technical challenges during implementation.

- o **Mitigation**: The solution architect and DevOps engineers will conduct a thorough review of the architecture and implementation plan.
- **Risk 2**: Performance issues with the NGINX deployment.
  - o **Mitigation**: The quality assurance engineer will conduct thorough performance testing and provide feedback for optimization.
- **Risk 3**: Downtime during deployment.
  - o **Mitigation**: The deployment will be scheduled during a maintenance window to minimize impact.

This project plan provides a roadmap for the successful deployment of a NGINX web server on an EKS cluster. The next step is to start with the planning and design phase and move towards successful deployment.

Architectural diagram

Refer to the following figure which shows an application running in AWS EKS however the concepts here would apply to any managed Kubernetes service, or any Kubernetes cluster for that matter:

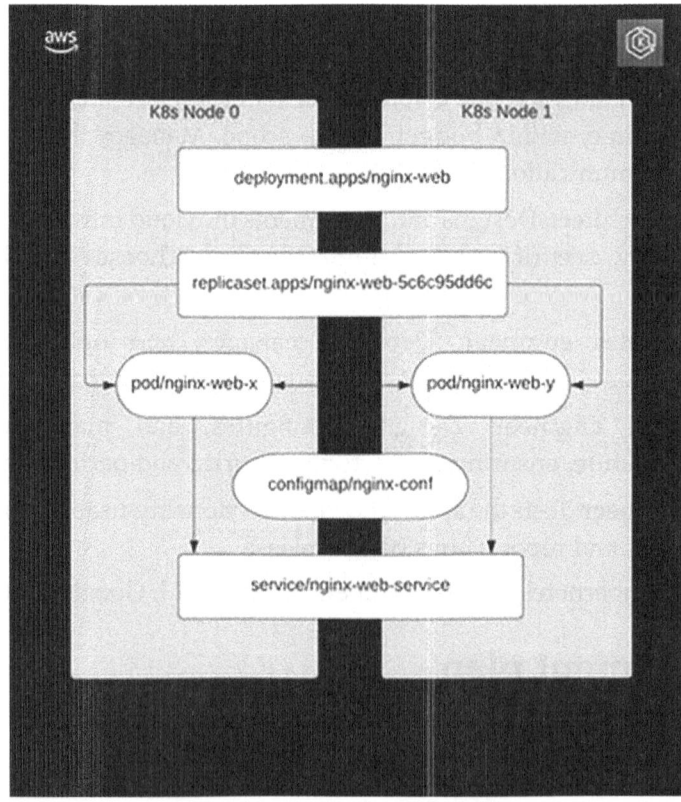

Figure 5.3: Architectural diagram

Implementation steps for deploying NGINX web server

We have our second opportunity to build a service into Kubernetes! For this exercise, we are going to continue to crawl before we walk or run. We are going to go through the steps of deploying NGINX web server on AWS EKS. In the coming chapters, we will build on this and add in the additional complexity of multi-cloud applications.

Prerequisites

The prerequisites are as follows:

- **Kubernetes clusters**: You should have an active Kubernetes cluster on AWS EKS. You should also have administrative access to these clusters.
- **Knowledge of Kubernetes**: A basic understanding of Kubernetes concepts such as pods, deployments, services, and ingress is required. You should also be comfortable with using kubectl, the command-line tool for Kubernetes.
- **Command Line Interface (CLI) tools**: You should have the AWS CLIand kubectl installed on your local machine or wherever you plan to manage your clusters from.
- **Docker**: Familiarity with Docker is essential as the applications are deployed within Docker containers.
- **Familiarity with YAML**: Kubernetes configurations are often written in YAML, so understanding how to read and write YAML files will be very helpful.
- **Networking knowledge**: Understanding of basic networking concepts, such as IP addressing, DNS, load balancers, and firewalls, is important as you will be configuring network resources.

Versions

This example assumes Kubernetes v1.26.

We will be creating the following resources to deploy NGINX:
- Namespace
- ConfigMap for NGINX web server configuration
- Deployment and ReplicaSet for NGINX web server
- Service to expose NGINX web server externally

We will first create the files, then apply them after in order. We will go into more detail on each resource before we create them.

Base configuration

A Kubernetes Namespace is a logical partition or grouping within a Kubernetes cluster that allows you to isolate and manage resources more effectively. Namespaces provide a scope for resource names, enabling you to have multiple instances of the same resource in different namespaces without conflicts. They can be used to divide cluster resources among multiple users, teams, or projects, and they help to maintain a clean and organized environment. In addition to resource separation, namespaces can be used to apply resource quotas and network policies, further enhancing resource management, access control, and security within a cluster.

Code:

```yaml
// namespace.yaml
apiVersion: v1
kind: Namespace
metadata:
  name: nginx-web
```

A Kubernetes ConfigMap is a resource used to store non-confidential configuration data in key-value pairs. It allows you to separate configuration information from the container image and the application code, making it easier to manage and update configurations independently. ConfigMaps enable you to decouple environment-specific configurations from your applications, which simplifies application deployment and scaling. Containers can consume ConfigMaps as environment variables, command-line arguments, or by mounting them as files in a volume. This provides a centralized and consistent way to manage configurations across multiple components in a Kubernetes cluster, making it more maintainable and flexible.

Code:

```yaml
// configmap.yaml
apiVersion: v1
kind: ConfigMap
metadata:
  name: nginx-conf
  namespace: nginx-web
data:
  nginx.conf: |
```

```
user nginx;
worker_processesauto;
error_log /var/log/nginx/error.log warn;
pid/var/run/nginx.pid;
events {
  worker_connections1024;
}
http {
  include /etc/nginx/mime.types;
  default_type application/octet-stream;
 log_format main '$remote_addr - $remote_user [$time_local] "$request" '
                  '$status $body_bytes_sent "$http_referer" '
                  '"$http_user_agent" "$http_x_forwarded_for"';
  access_log /var/log/nginx/access.log main;
  sendfileon;
  keepalive_timeout65;
  server {
    listen 80;
    server_namelocalhost;
    location / {
      root /usr/share/nginx/html;
      index index.html index.htm;
    }
    error_page 500 502 503 504 /50x.html;
    location = /50x.html {
      root /usr/share/nginx/html;
    }
```

    }
}

Deployment, ReplicaSet and Pods

A Deployment in Kubernetes is a high-level construct that provides declarative updates to applications. It allows you to describe an application's life cycle, such as which images to use for the app, the number of pod replicas, and the strategy to use when updating and rolling back to a new version. A Deployment is ideal for stateless applications where you need the flexibility of updating and maintaining app availability.

When you create a Deployment, it in turn creates a ReplicaSet. The ReplicaSet creates the Pods and ensures that the desired number of pods is always available and running. If a Pod fails, the ReplicaSet will create a new one to replace it and maintain the desired count.

Kubernetes ReplicaSet

A ReplicaSet's purpose is to maintain a stable set of replica Pods running at any given time. As such, it is often used to guarantee the availability of a specified number of identical Pods. A ReplicaSet creates and deletes Pods as needed to reach the desired state. If a Pod dies, the ReplicaSet will create a new one. If the ReplicaSet has too many Pods, it will kill the extras. A ReplicaSet is a lower-level concept and you do not usually need to manage it directly if you are using Deployments.

Kubernetes Pod

A Pod is the smallest and simplest unit in the Kubernetes object model that you create or deploy. A Pod represents a single instance of a running process in a cluster and can contain one or more containers. Containers within a Pod share an IP address and port space and can communicate with one another using localhost. They can also share storage volumes.

In summary, you create a Deployment when you want to run your application. The Deployment creates a ReplicaSet to ensure that the desired number of pods is always running. Each pod is an instance of your application running in the cluster.

Code:

```yaml
// deployment.yaml
apiVersion: apps/v1
kind: Deployment
metadata:
  name: nginx-web
```

```yaml
    namespace: nginx-web
spec:
  replicas: 3
  selector:
    matchLabels:
      app: nginx-web
  template:
    metadata:
      labels:
        app: nginx-web
    spec:
      containers:
      - name: nginx-web
        image: nginx:1.19.0
        ports:
        - containerPort: 80
        volumeMounts:
        - name: nginx-conf
          mountPath: /etc/nginx/nginx.conf
          subPath: nginx.conf
      volumes:
      - name: nginx-conf
        configMap:
          name: nginx-conf
```

Create the resources

When deploying a NGINX web server on Kubernetes, it is essential to create the resources in the following order to ensure proper configuration and avoid dependency issues:

- **Namespace**: Create the namespace to logically isolate and manage resources related to NGINX:

  ﹥kubectl apply -f namespace.yaml

  *Note: Set your context to your new namespace: "kubectl config set-context --current --namespace=nginx"
- **ConfigMap**: Create the ConfigMap containing the NGINX configuration settings. The ConfigMap should be created before the StatefulSet, as the StatefulSet references it for its configuration:

  ☐kubectl apply -f configmap.yaml
- **Deployment**: Create the deployment that deploys the NGINX instances. The deployment must be created after the ConfigMap as reference:

  ▢ kubectl apply -f deployment.yaml

Verify the NGINX web server deployment

After waiting a few minutes for all the resources to be created by the cloud provider, we can check that they are properly created and in the correct status:

﹥ k get configmap,po,replicaset,deploy

NAME	DATA	AGE
configmap/kube-root-ca.crt	1	15m
configmap/nginx-conf	1	7m43s

NAME	READY	STATUS	RESTARTS	AGE
pod/nginx-web-5c6c95dd6c-dlxrn	1/1	Running	0	7m34s
pod/nginx-web-5c6c95dd6c-g989c	1/1	Running	0	7m34s
pod/nginx-web-5c6c95dd6c-jp2j2	1/1	Running	0	7m34s

NAME	DESIRED	CURRENT	READY	AGE
replicaset.apps/nginx-web-5c6c95dd6c	3	3	3	7m34s

NAME	READY	UP-TO-DATE	AVAILABLE	AGE
deployment.apps/nginx-web	3/3	3	3	7m34s

Test the NGINX web server deployment

From your local machine, you can set up a port forward which will create a bridge between the NGINX pod and you. You set this up in terminal using the name of one of the pods. In this example we are using **pod/nginx-web-xxxx**, update it with your correct pod name:

```
>kubectl port-forward pod/nginx-web-xxxx 8080:80

Forwarding from 127.0.0.1:8080 -> 80

Forwarding from [::1]:8080 -> 80
```

Now you can make a call directly to the pod in another terminal window:

```
> curl -I http://localhost:8080

HTTP/2 200
```

Expose the NGINX web server

So now we have a functional and tested NGINX web server deployment in our EKS cluster. Now we want to expose the service so that it is available to services outside of the cluster. This is a crucial next step to setting up a multi-cloud deployment of our stateless web server!

To expose your web server deployment running in EKS to services outside of the cluster, you can create a Service with type **LoadBalancer**. This will expose the NGINX instance through a cloud provider's load balancer and assign a public IP address or DNS name to it.

Create a YAML file for the **LoadBalancer** Service:

```yaml
nginx-web-service-lb.yaml:

apiVersion: v1

kind: Service

metadata:
  name: nginx-web-service

spec:
  type: LoadBalancer
  ports:
  - port: 80
    targetPort: 80
```

```
    protocol: TCP
    name: http
  selector:
    app: nginx-web
```

Apply the Service YAML to your EKS cluster:

```
>kubectl apply -f nginx-web-service-lb.yaml
```

service/nginx-web-servicecreated

Get the public IP address or DNS name of the LoadBalancer created by the cloud provider for the Service:

```
>kubectl get svc -n nginx-web -o json | \
jq -r '.items[] | .status.loadBalancer.ingress[].hostname'
```

a541776d3b7f04f778210535a47c5104-1443515392.us-east-2.elb.amazonaws.com

You will see the public IP address or DNS name of the LoadBalancer.

Now, the NGINX web server is exposed and accessible from outside the cluster. You can use the public IP address or DNS name to connect to it from services running outside the EKS cluster.

Keep in mind that exposing NGINX over the public internet might pose security risks. In future chapters we will mitigate these risks in the following ways:

- Use encryption for data-in-transit by enabling SSL/TLS for the NGINX web server instances. You will need to configure the NGINX web server instances to use SSL/TLS and provide the necessary certificates and keys.
- Restrict access to the LoadBalancer by using security groups, firewall rules, or any other cloud-provider-specific access control mechanisms to only allow traffic from the IP addresses of the nodes or the services that require access to the NGINX instance.

Test out the public endpoint

We can test out the public endpoint in much the same way. Make sure the port-forward has been terminated (Control-c), then us W3M in the same way as before but using the public endpoint:

```
> w3m  a541776d3b7f04f778210535a47c5104-1443515392.us-east-2.elb.amazonaws.com
```

Refer to the following figure:

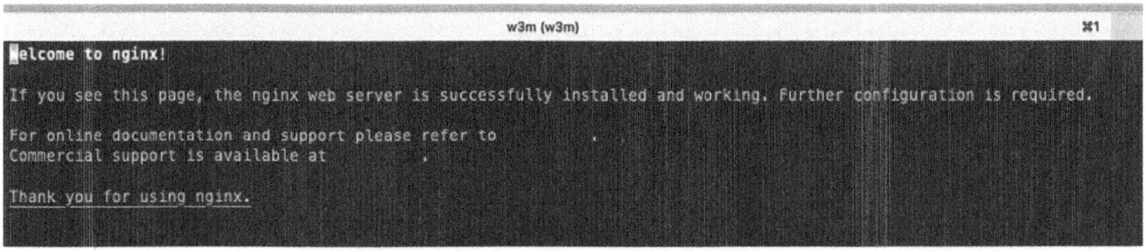

Figure 5.4: Correct response from working NGINX server

Conclusion

Congratulations on completing this chapter! You have made significant progress in understanding the intricacies of designing Kubernetes clusters tailored for both stateful and stateless applications. You have gained insights into different application types, their specific Kubernetes resource requirements, and how their design and management change when considering a multi-cloud environment.

This chapter introduced the paradigm of stateless applications and the unique challenges and opportunities they present in a multi-cloud environment. We discussed the architecture and design considerations specific to stateless applications, and you learned about load balancing, scaling, and deployment patterns. We walked through a detailed example of deploying a NGINX web server, a common stateless application, across EKS and GKE.

As we move forward, we will shift our focus to the operational aspects of multi-cloud Kubernetes environments. The next two chapters will specifically focus on the concept of service mesh. A service mesh is a configurable infrastructure layer for a microservices application, making communication between service instances flexible, reliable, and fast. It is implemented through a sidecar proxy for service instances, where the service mesh abstracts the network and the services from each other.

You will learn about its importance, how it works, and how to effectively leverage it in a multi-cloud environment. This will further enhance your ability to design, deploy, and manage robust, efficient, and resilient applications on multi-cloud Kubernetes clusters.

Your journey is far from over. On the contrary, it is just getting more exciting. So, stay tuned, keep exploring, and keep learning!

Join our book's Discord space

Join the book's Discord Workspace for Latest updates, Offers, Tech happenings around the world, New Release and Sessions with the Authors:

https://discord.bpbonline.com

CHAPTER 6
Service Mesh: Operations

Introduction

As we have transitioned from the design and setup phase to the operational phase of multi-cloud Kubernetes environments, one technology stands out as a cornerstone of modern, robust, and efficient operations, that is, the service mesh.

We will begin this chapter with an introduction to service mesh, exploring its purpose, and the operational benefits it brings to distributed systems like our multi-cloud Kubernetes setup. Knowing the history and the evolution of the modern service mesh helps us understand them better, and therefore, we will spend some time discussing that. We will then dissect the service mesh, understanding its core components, namely the control plane and the data plane.

Following this, we will introduce two pivotal technologies in the service mesh ecosystem, Istio, a powerful service mesh platform, and Envoy, a high-performance proxy that forms the backbone of the data plane.

We will then explore the operational capabilities of a service mesh, focusing on traffic control features such as routing, load balancing, service discovery, and failure handling. We will further talk about resilience patterns and how service mesh can aid in implementing timeouts, retries, and circuit breakers.

Observability is a critical aspect of operations, and we dedicate a section to understanding how service mesh can enhance tracing, metrics collection, and monitoring.

Structure

Here is an overview of the topics we will cover:

- Introduction to service mesh and its operational benefits
- History of service mesh
- Common use cases for service mesh
- Deciding when to implement a service mesh
- Understanding service mesh components: Control plane and data plane
- Deciding which control plane manager to use
- Data plane vendor comparison matrix
- Overview of Istio and Envoy
- Service mesh traffic control
- Implementing resiliency
- Observability in service mesh

Objectives

By the end of this chapter, you will have a comprehensive understanding of service mesh and its operational benefits. You will become familiar with its core components, the control plane and data plane, and you will have a solid understanding of two key technologies, that is, Istio and Envoy. You will learn how to use service mesh to control traffic, implement features such as routing, load balancing, service discovery, and failure handling. You will understand how to enhance the resiliency of your applications with timeouts, retries, and circuit breakers using service mesh. In addition, you will learn how service mesh can aid in observability through enhanced tracing, metrics, and monitoring. Finally, you will gain practical experience in deploying and operating a service mesh across multi-cloud Kubernetes clusters.

Introducing service mesh and its operational benefits

In modern distributed systems, applications often span multiple services that need to interact with each other. As this complexity grows, maintaining communication and enforcing policies between these services becomes increasingly challenging. Here is where a service mesh comes in.

A service mesh is an architectural pattern that aims to manage and control the service-to-service communication in a microservices architecture. It acts as an intermediary for inter-service communications and handles service discovery, load balancing, failure recovery, metrics, and monitoring. A service mesh also offers more complex operational functionality, such as A/B testing, canary deployments, rate limiting, access control, and end-to-end authentication. Refer to the following figure:

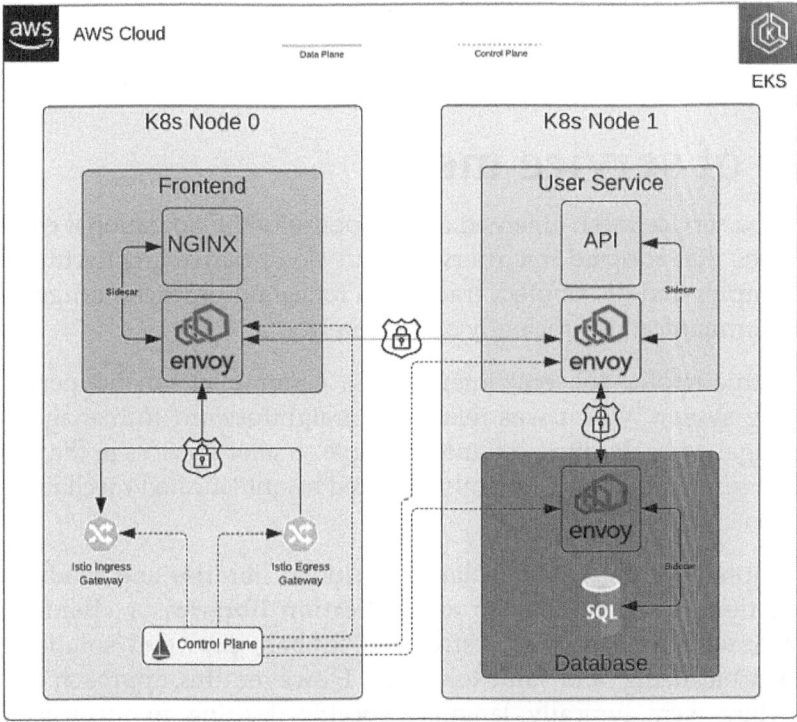

Figure 6.1: Service mesh traffic overview

One of the key operational benefits of a service mesh is that it abstracts the inter-service communication complexity away from the applications themselves. This separation of concerns allows developers to focus on building business logic, while operators can concentrate on managing the infrastructure.

With a service mesh, operators gain uniform, application-level networking capabilities that span the entire environment. This consistency simplifies the operational tasks and makes it easier to enforce security and compliance policies across all services. Moreover, the high observability provided by a service mesh is crucial for understanding the behavior and performance of microservices, which can aid in quickly identifying and resolving issues.

The next-generation traffic management capabilities of a service mesh, such as intelligent routing and circuit breaking, provide additional operational benefits. These features increase the reliability and resilience of the services, leading to enhanced application performance and a better end-user experience. Furthermore, these capabilities support modern development practices, like canary deployments, enabling safer and faster rollouts of new features.

In the multi-cloud Kubernetes environment, a service mesh can be especially valuable. It can provide consistent networking, security, and policy enforcement across clusters residing in different cloud environments, simplifying the operations, and providing greater control over the communication between services.

In the following sections, we will delve deeper into the components of a service mesh and explore how it achieves these benefits using platforms like Istio and Envoy.

History of service mesh

The concept of a service mesh emerged as a response to the operational challenges faced by organizations that adopted microservices and cloud-native architectures. As systems grew more complex and distributed, traditional tools and practices struggled to manage inter-service communication efficiently and securely.

In the monolithic application era, inter-process communication happened within the same operating system, which was relatively straightforward to manage. However, as applications began to be decomposed into multiple smaller services in the name of agility, scalability, and resilience, they inherently required a sophisticated mechanism to manage their interactions.

The first attempts to address this challenge included libraries and modules embedded within applications, often referred to as **application libraries** or **client-side libraries**. These libraries, such as Netflix's Hystrix and Ribbon, provided solutions for service discovery, load balancing, and fault tolerance. However, this approach had significant limitations. They were typically language-specific, leading to inconsistencies across different programming languages used in an organization. Furthermore, these libraries increased the complexity of the application codebase, making it harder to manage and evolve.

To address these issues, the industry began moving the communication logic out of the applications and into the infrastructure layer. This shift resulted in the **service mesh concept**, where a dedicated infrastructure layer handles service-to-service communications, freeing the services themselves to focus solely on business logic.

The initial service mesh solutions, such as *Buoyant's* Linkerd, were monolithic processes that handled all aspects of service communication. However, they added significant resource overhead and lacked some desired features, leading to the development of the modern, sidecar-based service mesh model.

The introduction of Envoy, a high-performance, C++ network proxy developed by Lyft, marked a significant milestone in the service mesh evolution. Envoy was designed from the ground up to be a self-contained sidecar proxy that could be deployed alongside any service, regardless of the programming language used. It provided rich features, including dynamic service discovery, load balancing, TLS termination, HTTP/2 and gRPC proxies, circuit breakers, health checks, and more.

Following Envoy's success, the Istio service mesh was launched, which used Envoy as its default data plane. Istio provided a robust control plane to manage a fleet of Envoy proxies and offered advanced traffic management capabilities, policy enforcement, and telemetry collection.

Today, service meshes have become a key component of cloud-native architectures, offering powerful operational benefits. By abstracting inter-service communication into a dedicated infrastructure layer, service meshes reduce complexity, increase developer productivity, enhance observability, and improve application security and resilience. As microservices continue to dominate the architectural landscape, the importance and role of service mesh are set to continue growing.

Common use cases for service mesh

Service meshes are becoming increasingly important in today's complex and highly dynamic application environments. They provide a level of abstraction that makes it easier for developers to build and deploy applications without worrying about the intricacies of inter-service communication. Here are some of the most common use cases for utilizing a service mesh:

- **Microservices observability and monitoring**: As applications grow and become more complex, it becomes increasingly challenging to observe and understand their behavior, especially in microservices architectures where many independent services interact. A service mesh provides detailed telemetry and tracing capabilities that offer deep insights into service-to-service communications. This helps to quickly diagnose issues, understand dependencies between services, and improve overall system performance.

- **Dynamic service discovery and load balancing**: In a microservices environment, services often scale up and down dynamically based on demand, and new versions are frequently deployed. A service mesh can automatically detect and register new service instances, allowing client services to discover them without manual intervention. They can also handle intelligent load balancing and traffic routing between services.

- **Network policy enforcement and security**: A service mesh can enforce fine-grained network policies, ensuring that only authorized services can communicate with each other. It can also handle encryption and decryption of service communication, providing automatic secure communication channels without the need to implement this in each service.

- **Resilience and fault tolerance**: Service meshes offer resilience features such as circuit breaking, retries, and timeouts out of the box, helping to increase the overall system's robustness. They can detect failures in the system and prevent them from cascading to other services, thereby improving the reliability of microservices architectures.

- **Traffic control and A/B testing**: With a service mesh, it is easier to manage traffic flow between services, implement canary deployments, and perform A/B testing. You can easily set up rules to direct a certain percentage of traffic to a new version of a service, enabling you to test new features or improvements in a controlled way.
- **Simplifying cross-cutting concerns**: Service mesh helps in standardizing and managing cross-cutting concerns such as logging, monitoring, security, and traffic management at the platform level, freeing developers from these responsibilities and allowing them to focus on business logic.

In summary, a service mesh is a crucial tool for managing, securing, and observing communication among services in a microservices architecture. These use cases are common in most modern application environments, making service mesh a vital component in the toolset for deploying and managing such systems.

Deciding when to implement a service mesh

Determining when to implement a service mesh can be a challenging decision. Service meshes are powerful tools that can provide immense benefits in complex environments, but they also bring additional complexity that may not be necessary or beneficial for every team or application. Let us discuss some considerations when deciding whether or not to adopt a service mesh.

You might need a service mesh when:

- **You are operating a microservices architecture:** A service mesh is most beneficial in a microservices environment where services need to communicate effectively and securely. The ability to provide consistent observability, resilience, and network policy enforcement across all services becomes a critical requirement as the number of services grows.
- **You need deep observability in your services:** When it is essential to have a detailed understanding of how services interact, a service mesh provides robust observability features such as monitoring, tracing, and logging.
- **You want to offload networking responsibilities from application code:** With a service mesh, developers can focus more on business logic rather than dealing with networking, security, and other related concerns.
- **You operate in a multi-cloud environment:** A service mesh can provide a uniform layer of infrastructure across multiple clouds, ensuring consistent policies, telemetry, and traffic control.

- **You need advanced traffic management capabilities:** If you require canary deployments, A/B testing, or blue-green deployments, a service mesh can simplify these tasks.

However, a service mesh may not be the right solution when:

- **Your application environment is relatively simple:** If you are managing a small number of services or a monolithic application, the additional complexity of a service mesh may not be worth the benefits.
- **Your team lacks the necessary skills or resources:** Implementing and managing a service mesh requires a certain level of expertise. If your team does not have the necessary skills or the time to learn them, it may be better to avoid a service mesh.
- **You are early in your cloud-native journey:** If you are just beginning to containerize applications or move to the cloud, it may be better to wait until these initial transitions are complete and stable before introducing the additional complexity of a service mesh.
- **Your network latency is a significant concern:** Service meshes can add a small amount of latency due to the additional hops required for service-to-service communication. While this is often negligible, it might be an issue for systems where every millisecond counts.

Remember, a service mesh is not a silver bullet, and it is not always the right solution. It is essential to understand your specific needs, challenges, and capabilities when deciding whether to implement a service mesh. Additionally, it is possible to introduce a service mesh gradually, starting with a small part of your application and expanding as necessary and beneficial.

Understanding service mesh components

A service mesh's architecture is commonly split into two key components: the Control Plane and the Data Plane. Each has a distinct role and together they provide the full feature set of a service mesh. Understanding these components is critical to comprehending how a service mesh functions and the benefits it provides.

Control plane

The control plane acts as the administrative interface for the service mesh. It is responsible for managing and controlling the behavior of the data plane. The control plane's core responsibilities include the following:

- **Configuration management:** The control plane accepts configuration changes from operators (for example, routing rules, security policies, and resilience settings) and propagates these changes to the data plane.

- **Policy enforcement**: The control plane enforces policies, such as authentication and authorization, rate limiting, and quotas, and applies them to the data plane.
- **Telemetry and reporting**: The control plane collects monitoring and telemetry data from the data plane, aggregates it, and presents it to operators for analysis.

Data plane

The data plane, often implemented as a sidecar proxy alongside each service, is responsible for the direct handling of network traffic. This includes tasks such as the following:

- **Traffic routing**: The data plane directs requests based on rules defined in the control plane. This might include directing traffic between different versions of a service for canary releases or load balancing requests.
- **Resilience handling**: The data plane implements resiliency features such as timeouts, retries, circuit breaking, and rate limiting, based on configurations from the control plane.
- **Security enforcement**: The data plane enforces security policies, such as **mutual Transport Layer Security** (**mTLS**), to secure service-to-service communication.
- **Observability**: The data plane collects telemetry data (metrics, traces, and logs) about requests and forwards it to the control plane.

In summary, the control plane and the data plane solve different problems and have distinct roles within a service mesh. The control plane manages the overall operation and configuration of the mesh, while the data plane handles the actual network traffic between services. The separation of these concerns allows each to be scaled and managed independently, enhancing the flexibility and robustness of the system.

Deciding which control plane manager to use

Choosing a service mesh control plane is a critical decision that can have a significant impact on your team's operational capabilities, productivity, and the stability and performance of your services. There are several key factors and requirements to consider when selecting a control plane, such as the following:

- **Features and capabilities**: You must first identify the features that are essential for your use case. This may include traffic control capabilities, policy enforcement, telemetry collection, service discovery, security enforcement, and more. Consider your use case, existing problems you are trying to solve, and the features that would be beneficial to your system.
- **Team experience and training**: The level of familiarity and comfort your team has with different control plane solutions is a significant factor. Some control planes may require more advanced knowledge and could have a steeper learning curve. The availability and quality of documentation and community support can also play a crucial role in easing this process.

- **Compatibility and integration**: The control plane must be compatible with your existing infrastructure and work seamlessly with your services. This includes compatibility with the networking, security, and observability tools already in use. It must also support your chosen container orchestrator, such as Kubernetes.
- **Performance and overhead**: The control plane should introduce minimal latency and have a low resource footprint. Evaluate the performance characteristics of the control plane under various workloads and scenarios to understand its impact on your services.
- **Scalability and high availability**: The control plane should scale to support the size and complexity of your service landscape. It should also be highly available and fault-tolerant to ensure continuous operation of your services.
- **Operational simplicity**: Managing a service mesh can be complex. Choose a control plane that simplifies operations as much as possible. This might include features for automated configuration, dynamic updates, and simple troubleshooting capabilities.
- **Community and support**: Consider the maturity of the control plane solution and the strength of its community and commercial support. This can help ensure you receive the necessary assistance and benefit from ongoing development and feature improvements.

By considering these factors, you can select a control plane that not only meets your immediate needs but also provides a solid foundation for future growth and complexity. It is also vital to pilot your chosen control plane in a non-production environment before fully committing to it, to ensure it fulfills your expectations and requirements.

Data plane vendor comparison matrix

A vendor comparison matrix is a tool used to identify the best solution or product from a list of vendors based on specific criteria and requirements. This matrix is typically structured in a tabular format where each row represents a critical factor or requirement, and each column represents a vendor or solution.

The matrix allows us to rate or measure each solution against various factors like features, pricing, support, integration, user-friendliness, and more. By comparing these solutions side by side, we get a clear visual representation of how each one fares in critical areas, helping us make an informed decision about which one best suits our needs.

In the context of selecting a service mesh control plane, a vendor comparison matrix can help us analyze how each solution measures up against important criteria.

In this vendor comparison matrix, we evaluate our requirements against to top 5 Control Planes, namely, Istio, Linkerd, Consul Connect, Kuma, Traefik Mesh.

Refer to the following table:

Factors and requirements	Istio	Linkerd	Consul Connect	Cilium	Traefik Mesh
Features and capabilities	High	High	Medium	High	Low
Team experience and training	High	Medium	Low	Low	Low
Compatibility and integration	High	High	High	High	Medium
Performance and overhead	Medium	High	Medium	Medium	High
Scalability and high availability	High	High	High	High	Medium
Operational simplicity	Medium	High	High	High	High
Community and support	High	High	High	Medium	Medium

Table 6.1: Control plane vendor comparison matrix

By using a vendor comparison matrix, we can more objectively assess each solution's strengths and weaknesses against our requirements, which can significantly aid in the decision-making process.

Deciding which data plane manager to use

The data plane is an integral part of a service mesh as it comprises the network components responsible for routing and forwarding data packets, facilitating the actual communication between services. Selecting the appropriate data plane to use involves carefully evaluating various factors that could significantly impact your service mesh implementation.

Here are some considerations to consider when deciding on a data plane:

- **Compatibility with control plane**: It is crucial to ensure that your data plane is compatible with your chosen control plane. For instance, Envoy is known for its excellent compatibility with Istio, and is actually the default data plane for it.
- **Performance**: This is a critical factor when selecting a data plane. The choice needs to have a minimal latency impact on service communication and must efficiently manage system resources.
- **Scalability**: As your application scales, your data plane will need to scale with it. Consider how easy it is to add new proxies to the data plane and how well the data plane handles large numbers of services.
- **Observability**: It is essential to monitor service communication for a variety of reasons, including troubleshooting and performance tuning. The data plane of choice should support robust metrics, logging, and tracing capabilities.

- **Security**: The data plane must provide essential security features, including mTLS for service-to-service communication, fine-grained access control, and protection from attacks like DDoS.
- **Resiliency**: The ability to handle failures gracefully is crucial for maintaining service reliability. The data plane should support features like automatic retries, circuit breaking, and rate limiting.
- **Protocol support**: Depending on the kind of workloads you run, you may need a data plane that supports HTTP/2, gRPC, or other protocols.
- **Team experience and training**: The level of familiarity and comfort your team has with different data plane solutions is a significant factor. Some data planes may require more advanced knowledge and could have a steeper learning curve. The availability and quality of documentation and community support can also play a crucial role in easing this process.
- **Community support and documentation**: An active community and detailed, up-to-date documentation can significantly simplify the implementation and troubleshooting process.

It is important to evaluate these based on the above criteria and your unique requirements. Moreover, as you noted, the existing skill set, and experience of your team can greatly influence this decision. If your team has already been working with a particular proxy and has a good grasp of it, it might be beneficial to choose it as your data plane. On the other hand, if the team is open to learning a new proxy and if it aligns better with your requirements, you might want to consider switching to a new data plane proxy.

Data plane vendor comparison matrix

We can do another vendor comparison matrix for the Data Plane. In this vendor comparison matrix, we evaluate our requirements against to top 3 Data Planes, namely, Envoy, Linkerd2-proxy, and Nginx.

Refer to the following table:

Criteria	Envoy	Linkerd2-proxy	Nginx
Compatibility with control plane	High	Medium	Low
Performance	High	High	Medium
Scalability	High	High	Medium
Observability	High	Medium	High
Security	High	High	Medium
Protocol support	High	High	High
Resiliency	High	High	Medium

Criteria	Envoy	Linkerd2-proxy	Nginx
Team experience and training	Medium	Low	High
Community support and documentation	High	Medium	High

Table 6.2: Data plane vendor comparison matrix

Please adjust the ratings based on your team's expertise and your specific use case requirements. Again, here we can take advantage of this process to aid in our decision-making process. It is a very powerful tool!

Overview of Istio and Envoy

After a careful comparison and discussion with stakeholders we have decided to use Istio and Envoy; we will be using these for the remainder of the book to illustrate examples as well as to build out a functional service mesh. Let us quickly review the pros and cons of Istio and Envoy as well as how they work together to form a reliable and robust service mesh.

Istio as a control plane

Istio, as an open-source service mesh, has emerged as a robust and feature-rich control plane solution that enables developers and operators to connect, secure, control, and observe services across clusters and clouds.

Pros of Istio

The pros of Istio are as follows:

- **Robust traffic management:** Istio's traffic routing rules allow for easy A/B testing, canary releases, and staged rollouts. It also supports load balancing for HTTP, gRPC, WebSocket, and TCP traffic.
- **Security:** Istio secures service-to-service communication at the network and application layers, providing identity and credential management, secure naming, and robust authorization policies.
- **Observability:** It provides a unified telemetry system with out-of-the-box integrations for logging, monitoring, and tracing, which reduces the need for additional libraries within your applications.
- **Multi-cluster and multi-cloud support:** Istio enables seamless traffic routing across Kubernetes clusters in different regions or cloud providers, making it an excellent choice for multi-region and multi-cloud deployments.

Cons of Istio

The cons of Istio are as follows:

- **Complexity:** Istio's feature richness comes with a degree of complexity that can be overwhelming for beginners.
- **Resource consumption:** Istio's control plane can be resource-intensive, which might affect its suitability for smaller clusters.

Envoy as a data plane

Envoy, an open-source edge and service proxy, serves as Istio's default data plane. As a service proxy, it mediates all network communication between microservices along with being responsible for service discovery, health checking, load balancing, authentication, and authorization.

Pros of Envoy

The pros of Envoy are as follows:

- **Performance**: Written in C++, Envoy is designed for high performance. It provides a concurrent, event-driven model that makes it highly efficient in terms of memory usage and CPU.
- **Flexibility**: Envoy supports dynamic configuration, hot restarts, and a pluggable filter architecture. This allows for dynamic responses to changing operational conditions.
- **Observability**: Envoy emits a plethora of statistics, providing detailed visibility into network behavior, latency, and performance.
- **Protocol support**: Envoy supports a wide range of protocols, including HTTP/1.1, HTTP/2, gRPC, TCP with TLS and SNI, etc.

Cons of Envoy

The cons of Envoy are as follows:

- **Learning curve:** While Envoy is powerful and flexible, it comes with a steep learning curve due to its complex configuration and tuning options.
- **Configuration complexity:** Envoy's rich feature set results in extensive configuration requirements, which can be a barrier for teams that are new to it.

How do they work together

In a service mesh like Istio and Envoy, the control plane (Istio) manages and configures the data plane or service proxy (Envoy). Istio gathers the state of the system, like service discovery data, traffic management rules, and access policies from its API objects and propagates that data to the data plane, the sidecar Envoys running in the application Pods. These Envoys then enforce the policies and rules on the actual network traffic. This setup ensures a clear separation of concerns: The Istio control plane focuses on the what (configuration), and the Envoy data plane focuses on the how (execution).

In a multi-cloud or multi-region context, Istio can be configured to run in a variety of topologies, either with a single shared control plane or with separate control planes for each cluster. Service discovery and routing can be set up across clusters and cloud environments, enabling a uniform service mesh that spans across your entire infrastructure. We will be exploring these in the following chapters as well as in the case study at the end of the book.

Service mesh traffic control

In the context of a service mesh, traffic control refers to the ability to direct, manage, and manipulate the flow of network requests within and between your services. The service mesh enables fine-grained control over traffic, providing the capability to shape traffic behavior for specific services, versions, and more. Traffic control is an integral aspect of service mesh operation, providing visibility, resilience, and security.

For a single-cluster Kubernetes environment, traffic control is fairly straightforward. The service mesh operates within a single cluster, managing traffic within that environment. It can direct traffic to appropriate services, apply policies, handle load balancing, perform health checks, implement failover strategies, and so on. It essentially orchestrates communication within a self-contained ecosystem.

In a multi-region Kubernetes environment, traffic control becomes more complex. Here, the service mesh must manage traffic not only within clusters but also between clusters that are located in different regions. This requires an additional layer of routing intelligence to ensure that traffic is sent to the right cluster in the right region, considering factors like latency, cost, and availability. The service mesh may also need to account for different regional compliance requirements.

When we consider multi-cloud Kubernetes environments, traffic control involves managing communication across clusters running in different cloud environments (like AWS EKS, Google Cloud GKE, and Azure AKS). In this scenario, the service mesh must be cloud-agnostic, able to seamlessly route traffic across different infrastructure providers. It should also be able to manage the complexity of different networking models, security constraints, and service discovery mechanisms of these cloud environments.

Irrespective of the environment, the service mesh should allow for fine-grained traffic control policies, like canary deployments, blue/green deployments, A/B testing, and circuit breaking. Additionally, traffic encryption via mutual TLS, access control, rate limiting, and observability features like tracing and metrics collection are critical components of traffic control in a service mesh.

Whether the environment is single-cluster, multi-region, or multi-cloud, the goal of traffic control in a service mesh remains the same, that is, to ensure reliable, secure, and efficient communication between services. The difference lies in the complexity and scope of the operations.

Let us dive into some of the key components of traffic control in the context of single cluster, multi-region clusters as well as multi-cloud cluster environments.

Routing and load balancing

Routing and load balancing are fundamental aspects of traffic management in a service mesh. These mechanisms work together to optimize the delivery of requests across the services in a distributed system, and can dramatically improve the efficiency, resilience, and overall performance of applications.

Routing

Routing in a service mesh refers to the ability to determine and control the path that requests take through the mesh. With sophisticated routing rules, it is possible to direct traffic based on a wide variety of criteria. For example, you can send requests to different versions of a service based on the content of the request, the identity of the requester, or even the current load on the services. This is key to implementing strategies like canary releases or A/B testing.

Load balancing

Load balancing, on the other hand, is the process of distributing network traffic across multiple servers or pods to ensure no single server bears too much demand. This not only helps in distributing the traffic but also helps in increasing application availability and responsiveness.

In a single-cluster Kubernetes environment, routing and load balancing work at the intra-cluster level. All requests within the service mesh are distributed among the services based on predefined rules and the current state of the services. Istio, for example, offers a rich set of routing features, including traffic shifting, fault injection, and request timeouts.

In a multi-region Kubernetes environment, routing and load balancing are responsible for distributing traffic across clusters in different regions. The service mesh should be able to route traffic to the appropriate region based on latency, cost, or other custom rules. Load balancing ensures that traffic is evenly distributed among available clusters, considering factors such as region capacity and demand.

In a multi-cloud Kubernetes environment, the challenge of routing and load balancing is compounded due to the inherent complexities of multiple cloud platforms. Here, the service mesh must seamlessly route and balance traffic across diverse infrastructure providers, each with their own networking models, security constraints, and service discovery mechanisms. The service mesh should be cloud-agnostic, with the ability to distribute traffic based on the performance and availability of services across different cloud platforms.

By efficiently managing routing and load balancing, a service mesh can help to ensure high availability, improve application performance, and enable effective deployment strategies across single cluster, multi-region, and multi-cloud Kubernetes environments.

Service discovery

Service discovery is an essential aspect of a service mesh, especially in distributed systems like Kubernetes, where applications and services are continually being scaled up and down. In such environments, the IP addresses of services are dynamic and can change frequently, making it challenging to manage network communication between services. This is where service discovery comes in.

Service discovery in a service mesh refers to the mechanism that allows services to find and communicate with each other without knowing their IP addresses. It enables services to register their presence when they become active and discover other services when needed, regardless of their location within the mesh.

In a single-cluster Kubernetes environment, service discovery is typically handled by Kubernetes itself. When a service is created in Kubernetes, it gets a stable network address that other services can use to communicate with it. Istio extends this model and uses Pilot, one of its core components, to provide service discovery for services within the mesh. Pilot converts high-level routing rules that control traffic behavior into Envoy-specific configurations and propagates them to the sidecars at runtime.

In a multi-region Kubernetes environment, service discovery becomes more complex. Services may need to communicate across different clusters, each potentially in a different network and geographic location. Istio supports multi-cluster service discovery through a variety of models, including single and multiple control plane models. Services are discovered and invoked by their **hostname**, which resolves correctly to the service's IP, regardless of the cluster or region it's located in.

In a multi-cloud Kubernetes environment, service discovery is even more complex due to differences in networking, security, and configuration between cloud providers. Again, Istio's Pilot component helps by abstracting the specifics of each platform and providing a unified service discovery interface. This means that services can find and communicate with each other, regardless of whether they're located in an on-premises data center, a Google Cloud cluster, or an Amazon EKS cluster.

By managing service discovery efficiently, a service mesh can facilitate seamless communication between services, improve network reliability, and enable effective load balancing across single cluster, multi-region, and multi-cloud Kubernetes environments.

Fault injection and traffic mirroring

Let us now learn more about fault injection and traffic mirroring.

Fault injection

Fault injection is a testing method used to increase the resilience of a system by intentionally introducing faults to validate how the system behaves under such conditions. In the context of a service mesh, fault injection is about deliberately introducing errors or delays into service communications to test the resilience of the services and the overall application.

Fault injection in a service mesh can take various forms, including introducing delays, aborting connections, or returning custom HTTP responses. This method can be used to test things like service retry logic, circuit breaking behavior, and overall system resiliency.

For single cluster, multi-region, and multi-cloud Kubernetes environments, fault injection behaves similarly, thanks to the abstraction provided by the service mesh. It allows developers to simulate errors and failures in any service, regardless of where it is located, and to observe the system's reaction. This helps in designing more robust and resilient applications.

Traffic mirroring

Traffic mirroring is another useful feature in a service mesh, which copies incoming traffic and sends it to a mirrored service. This mirrored service is often a newer version of the primary service or a service currently in testing or staging phase.

Traffic mirroring is extremely valuable when you want to test a new service or a new version of an existing service in a realistic environment with live data, without affecting the original service. This is also known as **shadowing**. You can then compare the performance and behavior of the new service against the existing service, allowing for safe testing and debugging.

Like fault injection, traffic mirroring in single cluster, multi-region, and multi-cloud Kubernetes environments is handled in a similar way by the service mesh. Istio, for instance, provides built-in support for traffic mirroring, which can be easily configured via its traffic management API. This means that developers can easily mirror traffic across clusters and clouds, which is particularly useful in a multi-cloud context where services may be spread across different cloud providers.

In essence, both fault injection and traffic mirroring provide developers and operators with powerful tools to improve system reliability and safety. They enable teams to proactively catch, handle, and learn from faults in the system and to test new services or service versions under realistic conditions, thus ensuring that the applications are ready for the complexities of production environments, regardless of their location.

Rate limiting and quota management

Let us now learn more about rate limiting and quota management.

Rate limiting

Rate limiting is a technique used to control the amount of incoming or outgoing traffic to or from a service. This can be extremely important for maintaining service quality, protecting services from abuse, preventing denial of service attacks, and ensuring fair usage among consumers.

In a service mesh like Istio, rate limiting can be configured to limit the number of requests a service can make or receive over a certain period. It can also be used to limit the data throughput. This functionality is particularly useful in microservices architectures where excessive or unwanted traffic to a particular service could cause it to slow down or even crash, potentially affecting the entire application.

Quota management

Quota management goes hand-in-hand with rate limiting and is about managing the overall resource usage by clients or services. It allows administrators to set limits on resources such as CPU, memory, and network bandwidth for individual services or clients. This is crucial for managing resources efficiently and ensuring that no single service or client consumes a disproportionate amount of resources.

In single cluster, multi-region, and multi-cloud Kubernetes environments, rate limiting and quota management are highly relevant and solve the same fundamental issues: preventing resource exhaustion and ensuring fair usage. They are instrumental in preventing any one service or tenant from overloading the system or causing disruptions, regardless of where the services are running.

In a multi-region and multi-cloud environment, these techniques help in managing traffic and resources across different geographical locations and cloud providers. For instance, rate limiting can prevent a sudden surge in requests in one region from affecting the performance of the system in another region. Similarly, quota management can ensure that resources are allocated fairly among services running on different cloud providers.

In short, rate limiting and quota management play a crucial role in maintaining the performance, reliability, and fairness of distributed systems in single cluster, multi-region, and multi-cloud environments. Their implementation in a service mesh provides a unified, flexible, and efficient mechanism for managing traffic and resources in a distributed system.

Access control and policy enforcement

Access control is a security measure that regulates who or what can view, use, or manage resources in a computing environment. In a service mesh, access control involves defining rules that allow or deny requests from one service to another. This plays a crucial role in protecting services from unauthorized access and data breaches.

Policy enforcement complements access control by ensuring all interactions between services comply with the rules and standards set by an organization. These rules can include everything from access controls, rate limiting, traffic routing, error retries, and more.

In a single cluster, multi-region, and multi-cloud Kubernetes environments, access control and policy enforcement solve several key issues such as the following:

- **Security and compliance**: By restricting who can access what services and enforcing policies uniformly across all services, a service mesh can greatly improve the security posture of the entire application and help it stay compliant with relevant regulations.
- **Consistent policy application**: In a multi-cloud environment, different cloud providers might have different mechanisms for implementing access control and policy enforcement. A service mesh abstracts these differences and provides a consistent layer for enforcing policies across all services, regardless of where they are running.
- **Micro-segmentation**: With the granular control provided by a service mesh, it is possible to implement micro-segmentation, isolating services and reducing the attack surface within a cluster or across multi-region and multi-cloud environments.
- **Reduced latency**: In a multi-region environment, policy enforcement at the service mesh level can reduce latency by routing requests to the nearest service instance that complies with the policy, rather than relying on a central policy management service.
- **Improved observability and audit**: With all policy enforcement and access control happening at the service mesh level, it is easier to monitor, trace, and audit these decisions, leading to improved transparency and accountability.

In conclusion, access control and policy enforcement at the service mesh level play a crucial role in maintaining security and compliance, enhancing performance, and improving observability and audit capabilities in single cluster, multi-region, and multi-cloud environments. By providing a consistent, flexible, and efficient mechanism for managing access and enforcing policies, a service mesh becomes an indispensable tool in managing complex, distributed systems.

TLS and mTLS

Transport Layer Security (TLS) is a cryptographic protocol designed to provide secure communication over a computer network. In the context of a service mesh, TLS is implemented to provide secure communication between the services in the mesh. This typically includes both TLS and mTLS for communication within the mesh.

In a single cluster, multi-region, and multi-cloud Kubernetes environments, TLS within a service mesh solves several key issues such as the following:

- **Secure communication**: The primary purpose of TLS is to provide secure, encrypted communication between services. In a single cluster or across multiple clusters and regions, ensuring that data in transit is protected from interception or tampering is crucial.
- **Identity verification (mTLS)**: In addition to encryption, mutual TLS provides a way to verify the identity of the services communicating with each other. This is especially important in multi-region and multi-cloud environments where services might be spread across different trust zones.
- **Consistent security posture**: In a multi-cloud scenario, each cloud provider might have different methods for implementing and managing TLS. A service mesh abstracts this complexity and provides a consistent way to implement and manage TLS across all services, regardless of where they are running.
- **Ease of management**: Managing TLS certificates can be complex and error-prone, especially when done manually for each service. A service mesh can automate the management of certificates, including issuing, renewing, and revoking them, making it easier and less prone to human error.
- **Improved performance**: By offloading TLS termination to the service mesh (specifically the data plane proxies like Envoy), you can potentially reduce the resource usage on your service instances and thereby improve overall performance.

By integrating TLS into a service mesh, you effectively secure service-to-service communication across the network, while simultaneously alleviating the headache of managing individual certificates for each service. This is a vital aspect of security, especially in complex single cluster, multi-region, and multi-cloud environments where communication routes can be highly variable.

Implementing resiliency

Resiliency in a service mesh refers to the ability of the mesh to tolerate failures and continue functioning in the face of partial outages or degraded services. The service mesh provides several powerful tools to help make your applications more resilient.

In a single cluster, these resiliency features are applied within the boundary of the cluster to ensure smooth operation of services within the cluster.

In multi-region deployments, resiliency becomes even more critical as network latencies, regional failures and other issues can impact the availability and performance of services. Service mesh can help ensure requests are automatically routed to available services in different regions if required, providing an additional layer of resiliency.

In multi-cloud environments, a service mesh provides uniformity in implementing resiliency across different cloud providers, each having their unique implementations of load balancing, network routing and so on. It abstracts the underlying infrastructure complexities, allowing you to focus on building and running your applications.

By integrating these resiliency features into the service mesh, you ensure that your applications can tolerate failures and continue to serve requests, improving overall system reliability and user satisfaction.

Let us dive into some of the key components of traffic control in the context of single cluster, multi-region clusters as well as multi-cloud cluster environments.

Retries

In a service mesh, **Retries** is a resiliency feature that is designed to tackle the issue of transient or intermittent failures within the network, or in the service itself. These transient failures could be due to a variety of reasons like network hiccups, overloaded services, temporary unavailability, and so on. The idea of retries is to attempt the same request again, assuming that the problem might resolve itself in the short term.

In a single-cluster environment, retries can be crucial for improving the reliability of the service requests within the mesh. For instance, if a particular service instance becomes temporarily unavailable due to a sudden spike in traffic or a brief network issue, retrying the failed request can help to ensure that the service's temporary unavailability does not result in a failed request.

In a multi-region deployment, retries become even more crucial. Network latencies, regional outages, or other inter-region communication issues can lead to temporary service unavailability. In such scenarios, retrying the failed requests can help maintain the application's availability and responsiveness. It is important to consider the increased latency for retries between regions.

In a multi-cloud environment, retries provide a way to handle intermittent failures across different cloud providers. Each cloud provider may have different levels of network reliability, and retries help ensure that a temporary issue in one cloud provider does not lead to a service disruption. Moreover, cross-cloud traffic might be less reliable or have higher latency, making retries an important part of request handling.

While retries can significantly enhance service availability, it is also important to be aware of the associated risks like retry storms (when large numbers of retries lead to overwhelming the system), increased latencies, and so on. Hence, effective strategies like exponential backoff, jitter, and maximum attempts should be employed to handle retries sensibly.

Overall, retries are a powerful tool that a service mesh provides to improve the resiliency of your microservices architecture. It's a crucial part of your strategy to ensure that temporary issues do not result in service disruptions or degraded user experiences.

Timeouts

Timeouts in a service mesh are essentially a resiliency mechanism that sets an upper bound on the amount of time a service call is allowed to take. If the request is not completed within the defined timeout period, it is considered as a failure and the client is informed accordingly.

In a single-cluster environment, timeouts are important to prevent a service from being indefinitely occupied by a request that is taking too long, which could be due to issues in the service itself, or the client making the request. This mechanism frees up the service to process other incoming requests and ensures that a single problematic request does not lead to resource exhaustion or degrade the overall performance of the service.

In a multi-region deployment, timeouts take on even more significance. Given that the services might be communicating across regions, network latencies and differences in the response times of services in different regions could potentially lead to long delays. With appropriately configured timeouts, such delays can be handled gracefully, ensuring that they do not propagate through the system and cause a domino effect of delays and failures.

In a multi-cloud environment, the role of timeouts is similar to that in a multi-region environment, but with additional complexity due to the heterogeneity of different cloud platforms. Cloud providers may have varying performance characteristics and service level agreements, and the use of timeouts allows for better control and predictability when dealing with these variations.

However, it is important to set sensible timeout values that balance the need for timely failure detection with giving requests enough time to complete under normal conditions. Moreover, timeouts should be used in conjunction with other resiliency features like retries and circuit breaking to prevent retry storms and to make the system more responsive to failure conditions.

In conclusion, timeouts are an essential aspect of implementing resiliency in a service mesh, offering a fail-fast mechanism that enables quick recovery from issues and contributes to maintaining the reliability and availability of services across single cluster, multi-region, and multi-cloud environments.

Circuit breakers

Circuit breakers are a key resiliency feature in a service mesh that help prevent a network or a service from being flooded with requests when it is experiencing errors. They operate similar to an electrical circuit breaker, when the number of consecutive failures crosses a defined threshold, the circuit breaker trips, and all further requests are automatically failed for a predetermined amount of time. Once the reset timeout is reached, the circuit breaker allows a limited number of test requests to pass through. If those requests succeed, the circuit breaker resets and allows normal traffic to resume. If the failures continue, the circuit breaker returns to the open state and the process repeats.

In a single-cluster environment, circuit breakers can prevent a faulty service from becoming a bottleneck, improving the system's overall ability to handle failures and recover from them.

In multi-region deployments, the use of circuit breakers is critical because they help prevent failures in one region from cascading into other regions. They are especially helpful when the services in different regions are interdependent.

In multi-cloud environments, circuit breakers provide an added level of protection as they prevent a problem in one cloud provider from spreading to services running on another provider. Each cloud platform may have its unique challenges, like throttling policies, network issues, or region-specific outages, and circuit breakers can help isolate these problems.

While circuit breakers significantly enhance the resiliency of a system, they also require careful configuration. They should not trip too easily, as it could cause unnecessary service denials, and they should not be too resistant to opening, as it might delay the detection of a real problem. Additionally, circuit breakers should be used in tandem with other strategies like retries and timeouts for comprehensive fault tolerance.

In conclusion, circuit breakers in a service mesh contribute to a resilient and self-healing system that maintains high availability, even when individual services or network paths fail. This resiliency is vital to service mesh deployments across single cluster, multi-region, and multi-cloud environments.

Health checks

Health checks are an integral part of implementing resiliency in a service mesh, as they monitor the state of services and determine whether they can handle requests. Health checks are typically of two types: liveness checks and readiness checks. Liveness checks determine if an instance is still running, while readiness checks assess if an instance is ready to accept requests.

In a single-cluster environment, health checks help detect issues within services running on individual pods. For instance, if a service fails a liveness check, Kubernetes can restart the offending pod, thus recovering from a potential application-level issue. If a service fails a readiness check, the service mesh can stop routing traffic to the failing instance until it recovers, which prevents requests from reaching a service that is unable to handle them properly.

In multi-region deployments, health checks are critical in ensuring high availability and reliability. When services span multiple regions, health checks can detect a regional failure, and the service mesh can route requests away from the affected region to another, helping maintain application availability despite regional outages.

In multi-cloud environments, health checks become even more crucial. They provide visibility into service health across different cloud providers, each with their unique challenges and potential points of failure. If a particular cloud provider experiences an outage, health checks allow the service mesh to reroute traffic to services running on a different provider, thus ensuring uninterrupted service availability.

While health checks play a crucial role in maintaining service availability, they also need careful configuration. The sensitivity of health checks needs to balance between not reacting to minor, transient errors, and promptly detecting major issues affecting service health. Overly sensitive health checks could result in unnecessary pod restarts or traffic rerouting, while checks that are not sensitive enough may result in failed requests reaching unhealthy services.

In conclusion, health checks in a service mesh enable continuous monitoring of service health, playing a crucial role in maintaining service availability and resilience. By promptly detecting issues and allowing for automatic recovery mechanisms, they significantly enhance the resilience of service mesh deployments across single cluster, multi-region, and multi-cloud environments.

Load balancing

Load balancing is a crucial aspect of implementing resiliency in a service mesh. At its core, load balancing involves evenly distributing network traffic across a group of backend servers, also known as a **pool of servers**. Load balancing aims to optimize resource use, maximize throughput, minimize response time, and avoid overloading any single resource.

In a single-cluster environment, load balancing is primarily concerned with distributing traffic amongst various pods within the cluster. This ensures that no single pod becomes a bottleneck, thereby optimizing resource utilization and improving the application's overall performance and availability. Additionally, should a pod fail, the service mesh can seamlessly redirect its traffic to other healthy pods, thus ensuring high availability and reliability.

In multi-region deployments, load balancing takes on an additional layer of complexity as traffic needs to be distributed not just across pods but also across different geographic regions. This could be based on a variety of factors such as the geographical proximity of the user to a particular region, the current load on different regions, and the health status of the regions. This kind of intelligent routing enhances the user experience (by reducing latency) and increases the resilience of the system by ensuring that a failure in one region does not impact the overall availability of the application.

In multi-cloud environments, load balancing must consider the unique characteristics of different cloud environments. In such scenarios, traffic may need to be distributed across different clusters hosted on different cloud platforms. This can provide an extra layer of redundancy and resilience, as issues in one cloud provider's infrastructure would not affect the services running on another provider's infrastructure.

However, load balancing across multiple clouds introduces additional challenges, like the varying performance characteristics, pricing models, and feature sets of different cloud providers. As such, a multi-cloud load balancing strategy would need to consider these differences to optimize cost and performance while ensuring high availability and resilience.

In summary, load balancing in a service mesh plays a critical role in optimizing resource use, enhancing performance, and ensuring high availability and resilience. By effectively distributing traffic across pods, regions, and cloud providers, it ensures that the system can effectively handle variable loads and recover swiftly from potential failures.

Rate limiting

Rate limiting is an important technique for maintaining the resilience and stability of applications within a service mesh. It is the act of limiting the number of requests a client can make to a service in a specific time period. The primary objective of rate limiting is to prevent resources from being exhausted by a high volume of requests, which can be caused by anything from a sudden surge in user traffic to malicious activity or even a simple misconfiguration.

In a single-cluster environment, rate limiting can prevent any single service or set of services from becoming overloaded and thereby becoming a bottleneck, potentially affecting the performance of the entire application. Rate limiting ensures fair usage of resources among services and can help in protecting services against **Distributed Denial of Service (DDoS)** attacks by limiting the number of requests a particular client can make.

In multi-region deployments, rate limiting can be more complex as it needs to consider the geographic dispersion of services. Depending on latency, cost, and regional regulations, it may be necessary to limit the rate of requests to or from certain regions. Effective rate limiting can ensure a balanced distribution of traffic and prevent any one region from becoming overwhelmed.

In a multi-cloud environment, rate limiting can be even more crucial, as different cloud providers may have different usage limits, pricing models, or network capacity. Furthermore, in these environments, traffic may be tunneled between clouds, potentially leading to high network costs or bottlenecks if traffic is not effectively managed. Rate limiting in this context can help control these costs and maintain a balanced distribution of traffic across clouds.

Overall, implementing rate limiting within a service mesh plays a key role in maintaining the application's resilience. By protecting against resource exhaustion, rate limiting helps ensure the stability and availability of services in the face of unpredictable traffic patterns or potential malicious activity. It is a critical tool in maintaining the overall health and performance of services within a service mesh, regardless of the environment's complexity.

Outlier detection

Outlier detection in the context of a service mesh is a method of identifying instances (pods in the case of Kubernetes) that are underperforming or behaving anomalously. It is a critical feature that contributes to the resilience of your services within the service mesh.

In a Kubernetes environment, whether it is a single cluster, multi-region, or multi-cloud, applications are often composed of multiple microservices. These microservices are typically replicated across multiple pods for availability and load balancing. With such a distributed system, certain instances can start behaving differently due to reasons such as resource contention, hardware failure, network latency, or software bugs.

Outlier detection, provided by the service mesh data plane (like Envoy in Istio), solves these problems by actively monitoring each instance's health based on configurable parameters like consecutive errors, latency, and so on. When an instance is deemed unhealthy or an outlier, it is ejected from the load balancing pool, ensuring that the service's overall performance and availability are not affected by this single misbehaving instance.

In a multi-region or multi-cloud environment, outlier detection becomes even more critical. Latency between regions or cloud providers can vary significantly and dynamically. Similarly, issues such as regional outages in a cloud provider, network partitioning, or rate-limiting can affect the services unevenly. By identifying and isolating the outliers, a service mesh ensures that the traffic is routed to the healthy and well-performing instances of the services, improving the application's overall responsiveness and availability.

It is important to note that outlier detection is a dynamic and automatic process that does not require manual intervention, making it a valuable tool in maintaining the resilience of your services in a complex multi-cloud Kubernetes environment.

Observability in service mesh

Observability, in the context of a service mesh, refers to the ability to monitor and understand the behavior of a system by inspecting its outputs. A highly observable system will provide clear insights into its performance, enabling you to understand what is happening within the system without needing to know the internal states. This is crucial for debugging, performance optimization, and general system health monitoring. Service meshes enhance observability through features such as tracing, metrics, and logging.

In a single-cluster environment, observability allows you to view the performance and status of all microservices in the cluster. With a service mesh like Istio, you can get a granular view of the system, including the success rate of individual service requests, the latency between services, and other critical information. This kind of insight can significantly improve your ability to detect and respond to issues rapidly.

In a multi-region deployment, observability becomes more complex and more critical. Services may span multiple regions, and network latencies, error rates, and other characteristics can vary from one region to another. The observability features of a service mesh can help identify regional anomalies, provide insights into user experiences in different regions, and offer data that can help in making decisions about resource allocation and load balancing.

In a multi-cloud environment, the importance of observability is further magnified. Different cloud platforms may have different performance characteristics and the services themselves may have dependencies that span multiple clouds. Here, a service mesh can provide uniform observability across all services, regardless of the cloud platform they are running on. This can simplify the task of managing and optimizing multi-cloud deployments and improve the overall robustness and resilience of the system.

Overall, observability is a key advantage provided by service mesh architectures. Whether you are working with a single Kubernetes cluster or a multi-cloud deployment, the enhanced observability provided by a service mesh can provide invaluable insights into system performance, improve your ability to debug and optimize your system, and ultimately lead to a more robust and reliable application.

Conclusion

In this chapter, we have embarked on an in-depth exploration of service mesh operations, taking significant strides in understanding the complex, multi-faceted world of multi-cloud Kubernetes. We started off by demystifying the concept of a service mesh, discussing its historical development, use-cases, and the situations when it is most appropriate to implement it. This gave us a foundation to understand the control and data plane components and how they cooperate to manage and route traffic within a service mesh.

We compared and analyzed various control planes and data planes, eventually choosing Istio and Envoy as our tools of choice for the upcoming parts of the book. A detailed overview of both tools showcased their strengths, features, and the unique ways they enhance enterprise-grade service mesh environments.

We then moved on to a comprehensive analysis of service mesh traffic control, discussing key components like routing, load balancing, service discovery, fault injection, traffic mirroring, rate limiting, quota management, access control, and policy enforcement. We also dove into the intricacies of TLS in a service mesh environment.

Our focus then shifted to resilience in a service mesh, with a detailed explanation of retries, timeouts, circuit breakers, health checks, load balancing, and rate limiting. Each of these elements contributes to a robust, resilient Service Mesh capable of withstanding a variety of scenarios.

The chapter ended with an exploration of observability, a critical element for understanding system performance and behaviors, particularly in multi-cloud Kubernetes environments.

As we venture further into the book, we will build a complete multi-cloud Kubernetes environment, leveraging the power of Istio and Envoy to form an enterprise-grade service mesh in our case study. As we transition into the next chapter, we will focus on the security aspects of service mesh, adding another layer of understanding to our multi-cloud Kubernetes journey.

Join our book's Discord space

Join the book's Discord Workspace for Latest updates, Offers, Tech happenings around the world, New Release and Sessions with the Authors:

https://discord.bpbonline.com

CHAPTER 7
Service Mesh: Security

Introduction

In this chapter, we turn our attention to an increasingly critical concern for businesses operating in digital space: Security. We delve into how service mesh architectures can reinforce security, with our chosen duo, that is, Istio and Envoy, playing pivotal roles in safeguarding our multi-cloud Kubernetes environment.

Structure

Here is an overview of the topics we will cover:
- Introduction to service mesh security
- Security principles in a service mesh
- Traffic encryption with mutual TLS
- Authorization and Access Control
- Network policy and isolation
- Security policies in control plane
- Threat modelling and security best practices in service mesh
- Security considerations

Objectives

By the end of this chapter, you will gain a solid understanding of service mesh security principles and mechanisms. We will explore in depth how **mutual TLS (mTLS)** secures service-to-service communication, and how authorization and network policies control access and restrict traffic within the mesh. You will learn about Istio's security policies and how to implement them, and we will discuss threat modeling and general security best practices in a service mesh environment. The final hands-on section will see you applying these concepts in a practical setting, enhancing security in a multi-cloud Kubernetes service mesh. It is our hope that with this knowledge, you can confidently and effectively secure your own service mesh deployments in any Kubernetes environment.

Introduction to service mesh security

Service mesh security has become an integral part of modern microservices architectures, especially with the emergence and proliferation of Kubernetes and container-based applications. Security, while always a critical concern, has gained further emphasis as applications become more distributed and complex.

The concept of a service mesh came into existence to simplify communication and control between different services in a distributed architecture. Initially, service mesh focused primarily on easing service discovery, fault tolerance, and load balancing. However, as more organizations moved towards microservices and faced increasingly sophisticated security threats, the need for robust security features within the service mesh became evident.

The evolution of service mesh security can be traced alongside the evolution of service meshes themselves. The first generation of service meshes provided basic communication security via network policies and firewalls, offering simple allow or deny rules for network connections. However, these features were rudimentary and lacked granular control. As service meshes evolved, security measures matured into fine-grained access control with traffic encryption, policy enforcement, authentication, and authorization.

Istio, for example, was introduced with a strong emphasis on security from the start. Leveraging Envoy's capabilities, Istio brought **mTLS** for service-to-service communications, offering both identity verification and encryption. It also introduced sophisticated access control policies, allowing for fine-grained control over who or what could access services within the mesh.

A comprehensive security posture is vital when implementing a service mesh. It allows for strong identity assertion, secure communication, and detailed control over who can do what within the mesh. This not only protects the services and data within the mesh but also provides observability and auditability, essential for compliance with various regulatory standards.

However, implementing security within a service mesh requires careful planning and consideration. Not only is it important to have a deep understanding of how your services work together, but it is also important to have a clear understanding of the security model provided by the service mesh, along with the potential risks and mitigations. This requires extensive knowledge of security best practices, familiarity with the specific service mesh technology (in our case, Istio and Envoy), and an understanding of the underlying Kubernetes platform. Training teams in these technologies and principles is a crucial step towards a successful and secure service mesh implementation.

Furthermore, it is essential to keep up-to-date with evolving industry standards and security practices, as the security landscape is always changing. Regular team training, attending security conferences, and following security-focused resources are good practices to ensure your organization maintains a robust, up-to-date security posture.

In summary, service mesh security has evolved significantly from its early days, and now provides a comprehensive suite of features to help secure your services and data. By understanding these features, staying current with best practices, and investing in team training, you can leverage these capabilities to secure your service mesh, regardless of whether you are operating in a single cluster, multi-region, or multi-cloud Kubernetes environment.

Security principles in a service mesh

Service meshes bring a paradigm shift to the way we handle the security of distributed systems. They offer several key security principles which make them a powerful tool for enforcing security in a microservices architecture. Let us discuss these principles and how they apply to service meshes.

Zero-trust network

In the context of a service mesh, the zero-trust network is a security model that assumes no inherent trust is given to anything or anyone inside or outside of the network. It operates under the principle, *never trust, always verify*.

Traditionally, security has often been perimeter-based, meaning that anything inside the network was trusted by default. However, this approach is fraught with risk, as once an attacker gains access to the network, they can move laterally with little to no resistance.

The zero-trust model takes a different approach and assumes that any network can be compromised. This ultimately results in strategies designed to minimize the impact of a breach by enforcing strict access control, limiting lateral movement, and providing thorough monitoring and logging for visibility and incident response.

Refer to the following figure:

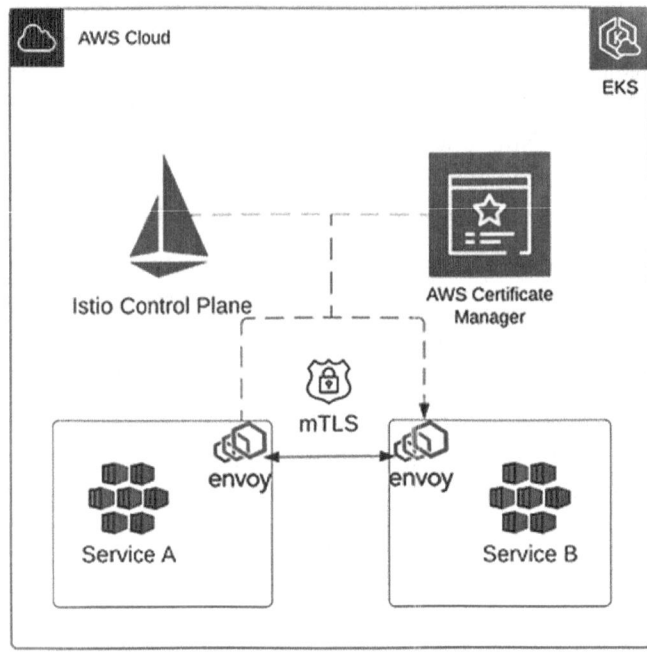

Figure 7.1: *Istio and Envoy*

In the case of a service mesh, this model is particularly relevant due to the increased complexity that comes with microservices architectures. Microservices are often spread across multiple networks, clouds, and possibly even geographical locations.

The application of zero-trust in a service mesh solves several problems such as the following:

- **Reduced attack surface**: By enforcing least privilege access at the microservice level, the potential attack surface is greatly reduced. An attacker gaining access to one service should not grant them access to others.
- **Improved visibility**: Zero-trust implies comprehensive monitoring and logging. In a service mesh, this can give a detailed overview of all communication between services, making it easier to spot anomalies and potential security threats.
- **Consistent policy enforcement**: With the help of a control plane like Istio, policies for traffic management and security can be uniformly applied across all services in the mesh, irrespective of where they are running - be it a single cluster, multiple regions, or across different cloud providers in a multi-cloud setup.
- **Secure communication**: Service meshes often employ mTLS for service-to-service communications, which not only encrypts communication for confidentiality but also ensures the identity of the communicating parties. This supports the zero-trust principle by ensuring secure, authenticated, and verified communication.

Remember that implementing zero-trust requires careful planning and understanding of your services and their communication patterns. It also requires clear policies and potentially significant configuration in your service mesh tools. However, the result is a more robust and secure application infrastructure that's better equipped to handle potential security threats.

Least privilege

Least privilege is a fundamental principle of information security that recommends that a user or application be given the minimum levels of access, or permissions, needed to perform its tasks. In the context of a service mesh, the principle of least privilege applies to how communications and interactions between services are managed and secured.

In a microservices architecture, as you would see in a service mesh, each service should have access only to the resources and other services that are necessary for it to perform its function. This concept becomes even more important in a distributed environment, such as a multi-cluster, multi-region, or multi-cloud Kubernetes deployment.

Implementing the principle of least privilege within a service mesh helps solve several security and operational problems:

- **Reduced attack surface**: By minimizing the permissions of each service, you limit the potential damage if a service is compromised. An attacker who compromises a service only has access to the resources that the service has access to.

- **Simplified audit and compliance**: With the least privilege, it is easier to audit the system and demonstrate compliance with various security standards and regulations. Since each service only has access to what it needs, you can quickly verify that no service has more access than it should.

- **Easier troubleshooting**: When services are operating under the principle of least privilege, it is simpler to identify issues and potential improvements. If a service does not have access to a necessary resource, the issue becomes evident when the service tries to perform a function that requires that resource.

- **Enhanced data protection**: When applied to data access, the principle of least privilege ensures that services can only access the data they need to function. This helps prevent unauthorized access and data leaks.

In a service mesh, least privilege is often implemented through the use of service identities and policies that define allowed communication paths. For example, a service mesh using Istio and Envoy can utilize **Role-Based Access Control (RBAC)** policies to define fine-grained access permissions for services. This approach works regardless of whether your service mesh is deployed in a single cluster, spans multiple regions, or is spread across multiple cloud providers in a multi-cloud setup. It is important to understand and map out the required communications between your services, in order to effectively implement the least privilege.

Mutual authentication

Mutual authentication, also known as **two-way authentication,** is a security process in which both entities in a communications exchange verify each other. This process ensures that each end of the communication is legitimate and can be trusted, providing a high level of security assurance.

In the context of a service mesh, mutual authentication typically involves verifying the identities of the services involved in a transaction. This is often achieved using cryptographic certificates that are used to prove the identity of each service.

The implementation of mutual authentication in a service mesh helps to solve several key problems such as the following:

- **Protection against impersonation and man-in-the-middle attacks**: With mutual authentication, both parties involved in a communication verify each other. This makes it extremely difficult for an attacker to impersonate a legitimate service or to position themselves in the middle of a communication to eavesdrop on or tamper with the data being exchanged.
- **Identity verification**: Mutual authentication helps to ensure that only legitimate, authorized services are communicating with each other. This is crucial for enforcing access controls and permissions within the service mesh.
- **Trust establishment**: Mutual authentication helps to establish a baseline level of trust in the service mesh. Services can trust that they are communicating with the intended services, and not with an impersonator.

In a single cluster, multi-region, or multi-cloud Kubernetes environment, implementing mutual authentication can add an extra layer of security. The distributed nature of these environments means there are more potential points of attack, making authentication and trust even more important.

In a multi-cloud environment, mutual authentication can be a bit more challenging, due to potential differences in security infrastructure across cloud providers. However, a service mesh like Istio provides capabilities such as Citadel for certificate and key management, and Envoy sidecars can handle the details of establishing mTLS connections, making it easier to implement mutual authentication consistently across diverse environments.

Mutual authentication is an essential security practice within a service mesh, and it should be part of the design of any service mesh, regardless of the scale or distribution of the services involved.

Secure communication

Secure communication is a fundamental security principle that ensures all communication between services in a service mesh is confidential, authenticated, and cannot be tampered with. It is typically achieved through the use of encryption protocols such as **Transport**

Layer Security (TLS) or its predecessor, Secure Sockets Layer (SSL).

In a service mesh, secure communication solves several key problems such as the following:
- **Confidentiality**: Encryption ensures that data in transit between services cannot be read if intercepted, preserving the confidentiality of the data.
- **Authentication**: In addition to encrypting data, protocols like TLS also provide mechanisms for authentication. This allows the services involved in communication to verify each other's identities, reducing the risk of attacks where an attacker might try to impersonate a legitimate service.
- **Integrity**: Secure communication protocols include mechanisms to ensure that data cannot be tampered with in transit. This provides assurances about the integrity of the data.

In single cluster, multi-region, or multi-cloud Kubernetes environments, the importance of secure communication is amplified. With services potentially distributed across different nodes, regions, or even cloud providers, the likelihood of data passing through potentially insecure networks is increased.

In a multi-cloud environment, it can be challenging to implement secure communication due to the potential differences in security infrastructure and certificate management across cloud providers. A service mesh, however, abstracts these details away and provides a unified way to achieve secure communication.

For instance, a service mesh like Istio provides automatic mTLS for all traffic within the mesh. The mTLS ensures both encryption and mutual authentication between services. Istio manages all the certificate issuance and rotation, providing secure communication with minimal configuration.

Therefore, secure communication is a crucial aspect of service mesh security, providing confidentiality, integrity, and authentication for service-to-service communications, regardless of the complexity or distribution of the mesh.

Fine-grained access control

Fine-grained access control is a core security principle for any application, not just within a service mesh. It refers to the ability to provide highly specific and detailed permissions, ensuring that components have the least privileges required to perform their functions. This level of access control significantly reduces the potential for malicious or inadvertent actions that can compromise the system.

In the context of a service mesh, fine-grained access control solves several problems such as the following:
- **Reduced blast radius**: If a component in the service mesh becomes compromised, fine-grained access control limits what that component can do, reducing the potential harm that can come from a security breach.

- **Enforce principle of least privilege**: Fine-grained access controls allow for enforcement of the least privilege principle. This principle states that a component should have only the privileges necessary to perform its function and no more. This minimizes the risk of privilege escalation.
- **Supports regulatory compliance**: Many industries have strict regulations for data access. Fine-grained access control can help meet these regulatory requirements by precisely controlling who or what can access sensitive information.

In a single-cluster, multi-region, or multi-cloud Kubernetes environments, fine-grained access control is crucial. In a single cluster, you can use Namespace and Network Policies to control traffic at a micro-level, defining which pods can communicate with each other.

In multi-region and multi-cloud environments, the complexity increases as different clusters in different regions or clouds may have different security requirements and regulations. A service mesh can abstract these complexities, enabling unified, fine-grained access control policies to be applied across all services, irrespective of where they are deployed.

For example, with Istio, you can implement fine-grained access control using authorization policies. These policies support a wide variety of selectors based on service identity, request headers, and more, allowing a high degree of specificity in defining who can do what in the system. By implementing fine-grained access control, you can significantly enhance the security posture of your service mesh in any environment.

Policy enforcement

Policy enforcement is an essential security principle in a service mesh and indeed for any distributed system. It is the mechanism by which the security rules and access controls are applied consistently across the mesh. Without policy enforcement, the security principles and rules we define would just be suggestions rather than enforced behavior.

Here is how policy enforcement within a service mesh addresses several challenges:

- **Consistency**: In a distributed system like a service mesh, maintaining consistency of policy enforcement can be challenging. Having policy enforcement built into the service mesh ensures that the same policies are applied uniformly across all services.
- **Speed and scalability**: The distributed nature of a service mesh means that policy decisions need to be made rapidly and at a scale. The policy enforcement mechanisms within a service mesh are designed to handle this requirement.
- **Reduced error and risk**: Manual policy enforcement is prone to error and can lead to security risks. In a service mesh, policy enforcement is automated, reducing the potential for human error and enhancing security.

Here is how this applies to different Kubernetes environments:
- **Single cluster:** Within a single Kubernetes cluster, policy enforcement can be relatively straightforward. Policies can be applied at the cluster level and enforced uniformly across all services within the cluster.
- **Multi-region:** When a Kubernetes deployment spans multiple regions, policy enforcement becomes more complex. Policies need to be consistent across regions, but there may also be region-specific requirements that need to be addressed. A service mesh can manage these complexities, enabling uniform policy enforcement while accommodating regional variations.
- **Multi-cloud:** In a multi-cloud environment, the complexities are even greater. Different cloud providers may have different security mechanisms and policies. A service mesh abstracts these differences, enabling consistent policy enforcement across all clouds in the deployment.

For example, Istio has a robust policy enforcement system. It uses a flexible policy language that allows the definition of complex rules and applies these rules consistently across all services in the mesh. These policies can enforce access controls, rate limits, and other behaviors that enhance security.

Observability

Observability, in the context of security principles in a service mesh, is a key aspect of maintaining a strong security posture. It is the practice of collecting, analyzing, and using telemetry data (like metrics, logs, and traces) to understand the internal state of your system and ensure security protocols are functioning as expected.

Here is how observability addresses several security challenges:
- **Threat detection**: Observability helps you quickly detect unusual or suspicious activity that might indicate a security threat, such as abnormal network traffic or unexpected API calls.
- **Incident response**: When a security incident does occur, observability data can provide the context needed to understand what happened and how to respond effectively.
- **Forensic analysis**: Observability tools can help record detailed activity data, which can be crucial for conducting forensic analysis after a security incident.
- : Observability can highlight where security controls might be impacting system performance and identify opportunities for optimization.

In terms of Kubernetes environments:
- **Single cluster**: In a single cluster, observability data can provide insights into interactions between services within the cluster and detect any security incidents within that environment.

- **Multi-region**: In multi-region deployments, observability becomes even more important. It can help ensure that security controls are consistently applied and effective across all regions.
- **Multi-cloud**: In a multi-cloud scenario, observability can help manage the complexity and variability of different cloud environments. It provides a unified view of security across all the services deployed in different clouds.

For instance, Istio and Envoy, which we have chosen as our service mesh components, provide robust observability features. Istio's control plane is able to log every interaction, track metrics of every service, and trace request flows. Envoy proxies can generate detailed access logs showing all traffic passing through the proxies, provide insights about request-level gRPC traffic, and more. This significantly helps in maintaining the security of your service mesh.

Secure ingress/egress

In the context of a service mesh, Secure Ingress and Egress involve controlling and securing the inbound and outbound traffic to your service mesh. It is an essential principle that enhances the security posture of your mesh by dictating what services can interact with your service mesh, either from inside or outside, and in what manner.

Let us look at how Secure Ingress/Egress addresses several problems:

- **Prevention of unauthorized access**: Secure Ingress controls ensure that only authorized traffic can enter your service mesh, preventing unauthorized users and potential attackers from reaching your services.
- **Data exfiltration control**: Secure Egress controls can help to prevent data exfiltration, which is a common way that attackers extract stolen data from a network.
- **Regulatory compliance**: Some regulations require specific controls over inbound and outbound network traffic. Secure Ingress/Egress helps meet these regulatory requirements.
- **Contextual access**: It allows for the use of more complex rulesets that evaluate the context of a request, not just its source and destination.

For Kubernetes environments:

- **Single cluster**: In a single Kubernetes cluster, secure Ingress/Egress controls can be implemented to manage the flow of traffic entering and leaving the cluster.
- **Multi-region**: In multi-region deployments, secure Ingress/Egress controls need to account for the additional complexity of routing traffic between regions. Policies should be synchronized and enforced across all regions to ensure consistent security.

- **Multi-cloud**: In a multi-cloud scenario, traffic may ingress or egress through various cloud providers, each potentially with different security controls and requirements. Secure Ingress/Egress policies must be robust and flexible enough to ensure consistent security across all clouds.

With Istio, for example, you can use Gateway configurations for Ingress/Egress traffic, allowing you to define fine-grained access control rules for inbound and outbound traffic. Together with Istio's Authentication and Authorization policies, it provides a powerful security model for your service mesh, regardless of whether you are running in a single cluster, multi-region, or multi-cloud environment.

Auditability

In the context of a service mesh, auditability refers to the ability to monitor, record, and analyze actions and changes within the system, especially those that may impact its security. This principle is essential in determining who did what, when, and where, as well as understanding how events may impact other components in the system.

Here is how auditability addresses several challenges:
- **Incident response**: In the event of a security incident, having an audit log helps incident responders understand the sequence of events leading to the incident, the extent of the breach, and the potential impact on other services.
- **Forensics and postmortem analysis**: In the aftermath of a security incident, audit logs serve as an important resource for forensic investigators and postmortem analysis. They can provide insight into the methods used by an attacker and help identify the source of a compromise.
- **Compliance and regulatory requirements**: Many industries and regulations require retaining audit logs for a certain period. These logs are essential for demonstrating that your system has the appropriate security controls in place and can assist with demonstrating compliance to auditors.

In different Kubernetes environments:
- **Single cluster**: In a single cluster environment, the service mesh's audit logs can provide valuable information about the communication and behavior of the microservices within the cluster. Istio, for instance, has the ability to produce access logs detailing all incoming and outgoing traffic to each service.
- **Multi-region**: In a multi-region setup, audit logs from each region can provide insights into the interaction between services across different regions. It is important that these logs are consolidated in a central location to enable comprehensive monitoring and analysis.
- **Multi-cloud**: In a multi-cloud environment, the complexity of auditability increases due to the diverse nature of cloud services and their individual logging mechanisms. However, the importance of auditability also grows with the

increased potential for security incidents. Ensuring consistent logging across all cloud providers and centralizing the collection of these logs for analysis becomes essential.

By adhering to the auditability principle in a service mesh, organizations can bolster their security posture, enhance incident response, and meet regulatory compliance requirements.

Remember that the application of these principles varies depending on the context of your environment, whether it is single cluster, multi-region, or multi-cloud Kubernetes. Understanding these principles will guide you in setting up and operating a secure service mesh, leveraging tools like Istio and Envoy.

Traffic encryption with mutual TLS

In the realm of service mesh security, mTLS plays a critical role. It not only protects the communication channels among the services inside the mesh but also validates the identity of the services, thus implementing secure and authenticated communication.

Mutual TLS

Mutual TLS (mTLS) is an extension of the **Transport Layer Security** (**TLS**) protocol, which is widely used to encrypt web traffic. mTLS enhances standard TLS by providing two-way authentication. This means both the client and server validate each other's identities before the transmission of data takes place. This contrasts with traditional TLS, where only the server's identity is validated.

Implementing mTLS in a service mesh

Implementing mTLS in a service mesh offers several advantages:

- **Identity verification**: mTLS uses certificates to verify the identity of client and server, providing a more secure method of identity verification than other methods like API keys or JWT tokens.
- **Data protection**: All data transmitted between services is encrypted, preventing unauthorized access to the data in transit.
- **Reduced complexity for developers**: In a service mesh like Istio, mTLS can be managed and enforced without changes to the application code, freeing developers from the need to implement TLS in each service manually.

The implementation of mTLS varies depending on the environment:

- **Single cluster**: In a single cluster environment, mTLS can be enforced at the mesh level. All services in the mesh will then communicate using mTLS, encrypting all the traffic.

- **Multi-region**: In a multi-region setup, mTLS can be used to secure communication between services across different regions. The service mesh control plane synchronizes the certificates and keys across regions.
- **Multi-cloud**: In a multi-cloud setup, maintaining a common root of trust for mTLS across different cloud providers can be a challenge. The service mesh control plane needs to synchronize certificates and keys across multiple clouds.

mTLS offers a strong foundation for secure communication within a service mesh, ensuring that sensitive data is not exposed during transmission and that the identity of the communicating parties is verified. In the next section, we will delve deeper into how mTLS can be configured and managed within the context of our chosen service mesh, Istio.

Authorization and access control

Authorization and access control are two fundamental aspects of service mesh security. They define who can access which resources and services within the mesh, and the operations they can perform.

Authorization

In the context of a service mesh, authorization refers to defining and enforcing policies that dictate which services (or users) can access other services and what operations they can carry out. For example, a policy might restrict service A from accessing service B, or it may allow service A to only read data from service B but not modify it.

Istio, the service mesh that we are using for the rest of this book, implements RBAC for services in the mesh. With RBAC, you can define roles that encapsulate a set of permissions, and then assign these roles to services. This way, each service in the mesh has a defined set of capabilities, limiting the potential damage if a service is compromised.

Access control

Access control involves implementing the authorization policies. It is the enforcement layer that ensures the policies are respected. Istio uses a combination of Envoy proxy and Mixer to enforce access control. Each service in the mesh has an associated Envoy proxy that intercepts all incoming and outgoing traffic. When a service tries to access another service, the Envoy proxy checks with mixer to see if the request is allowed under the defined policies.

The implementation of authorization and access control varies slightly depending on the environment:

- **Single cluster**: In a single cluster, Istio can enforce uniform policies across all services.

- **Multi-region**: For a multi-region setup, Istio can enforce region-specific policies, allowing for finer-grained control.
- **Multi-cloud**: In a multi-cloud scenario, separate Istio deployments in each cloud can manage the authorization and access control. A central Istio control plane can also synchronize policies across clouds.

In conclusion, authorization and access control are powerful security mechanisms in a service mesh. They help to protect your mesh from internal threats and limit the potential blast radius of compromised services. In the next section, we will explore the concept of network policy in the context of a service mesh.

Network policy and isolation

Network policy and isolation are crucial concepts in the realm of service mesh security. They primarily cater to the requirement of securing applications and their communication channels within the mesh.

Network policy

Network policies in the context of service mesh and Kubernetes act as a firewall between pods running in a cluster. They use labels to select pods and define rules which specify what traffic is allowed to the selected pods. This traffic could be ingress (coming to the pod), egress (leaving the pod), or both.

Istio's implementation of network policy enables fine-grained control over microservices, controlling who can talk to whom at the protocol level. Network policies can control traffic based on the source and destination protocol, ports, and source and destination labels, thus enhancing security and reducing the attack surface.

Isolation

Isolation, on the other hand, refers to limiting the visibility and interaction between different components or services within a service mesh. By isolating services, you reduce the 'blast radius' if a service is compromised. For instance, an attacker compromising one service will not automatically gain access to all other services in the mesh.

Istio provides namespace-level isolation. You can define a policy that only allows a service to access services in the same namespace, effectively isolating the namespace from the rest of the mesh. Isolation can be implemented at varying levels based on the environment:

- **Single cluster**: Within a single cluster, namespaces can provide a natural boundary for isolation. Services can be designed to communicate primarily within the same namespace, reducing cross-namespace traffic.

- **Multi-region**: In a multi-region scenario, isolation can be implemented at the region level. Services in one region can be prevented from communicating directly with services in another region.
- **Multi-cloud**: In a multi-cloud setup, isolation can be implemented at the cloud boundary. Services in one cloud provider can be prevented from communicating directly with services in another provider.

By incorporating network policy and isolation in your service mesh, you can significantly enhance the security of your applications, protect sensitive data, and reduce the risk and impact of breaches.

Security policies in a control plane

A control plane in a service mesh, such as Istio, manages the network traffic between the different microservices. It imposes a range of rules and policies that ensure the security, reliability, and observability of your applications. In terms of security, the control plane is responsible for implementing a range of security policies.

Here are the key security policies you can enforce through the control plane:

- **mTLS policies**: mTLS is a security protocol that ensures secure, authenticated, and encrypted communication between microservices in your service mesh. It is called mutual because both the client and server verify each other's identities before initiating communication.
- **Secure communication**: In a service mesh, especially in a multi-cloud Kubernetes environment, microservices could be distributed across multiple nodes, clusters, or even regions. This geographical distribution means the network traffic between these services has to traverse various networks, which could be vulnerable to attacks. mTLS policies enforce encryptions for all in-transit data, ensuring the communication remains confidential and immune to tampering or eavesdropping.
- **Service authentication**: mTLS goes a step beyond conventional TLS by requiring both the client and server to authenticate each other. This two-way authentication mitigates the risk of impersonation attacks. By ensuring that a service is communicating with a legitimate partner, mTLS prevents scenarios where an attacker might pose as a legitimate service.
- **Flexible trust domain validation**: With mTLS, you can implement different levels of trust validations across your mesh. For instance, in a multi-cloud environment, you might want to enforce stricter mTLS policies for traffic crossing cloud boundaries compared to intra-cloud communication. These flexible policies allow for enhanced security posture and fine-grained control.
- **Support for zero-trust security**: mTLS is a fundamental element of a zero-trust security model, where no entity is trusted by default and authentication is required from all components. mTLS ensures that every service in the mesh proves its

identity, thereby eliminating the trust typically given to components within the internal network.

In a nutshell, mTLS policies enforced by the control plane enhance the security of your service mesh by ensuring secure and authenticated communication between services. This is true whether the services are located within the same cluster, spread across multiple regions, or distributed across multi-cloud environments.

Authorization policies

Authorization policies in a service mesh control plane are designed to address the problem of ensuring that services and users have the appropriate permissions to perform certain actions within the system. In the context of a service mesh, authorization is particularly significant due to the complex interplay of multiple services, each possibly having different access needs and restrictions.

Here is a deeper look at the issues that authorization policies address in a control plane:

- **Fine-grained access control**: One of the primary challenges in any distributed system is managing who can access what. As the number of services grows, especially in multi-cluster and multi-cloud Kubernetes environments, it becomes critical to ensure only authorized services can access certain resources. Authorization policies in a control plane allow you to establish fine-grained access control rules, permitting specific services to interact while blocking others based on their identities or roles.

- **Policy enforcement and compliance**: Organizations often have stringent rules around data access to comply with regulations and business requirements. Authorization policies in the control plane ensure that these rules are enforced across all services in the mesh, irrespective of where they are deployed. This is especially critical in multi-region and multi-cloud environments, where data sovereignty and cross-border data transfer regulations might come into play.

- **Mitigation of insider threats**: While encryption and mTLS can secure your service mesh from external threats, insider threats remain a significant concern. A service or a user with more permissions than necessary poses a security risk. Authorization policies provide the mechanism to implement the principle of least privilege, thereby reducing the risk from overly permissive access rights.

- **Support for zero trust security model**: Just as with mTLS, authorization policies are a key component of a zero-trust security model, ensuring that every request is thoroughly checked for permissions. The authorization policies deny all access by default and only permit actions explicitly allowed, thereby strengthening the security posture.

In summary, by allowing precise control over who can access what, authorization policies in a control plane add an important layer of security in your service mesh. They provide a systematic way to control service interactions, enforce compliance requirements, mitigate insider threats, and implement a zero-trust security model, regardless of whether your services are deployed in a single cluster, across multiple regions, or in a multi-cloud Kubernetes setup.

Audit logging policies

Audit logging policies in a service mesh control plane tackle the challenge of monitoring, recording, and reviewing the operations and transactions happening within the mesh. Audit logs are critical for maintaining visibility and traceability of actions, which is key for both security and compliance requirements.

Here is a closer look at the issues that audit logging policies address in a control plane:

- **Tracking and accountability**: With services frequently interacting with each other, understanding who did what and when becomes a crucial requirement. Audit logging policies provide an ability to record all the actions taken within the service mesh, which can later be reviewed for inconsistencies or breaches. This is particularly valuable in multi-cluster and multi-cloud Kubernetes environments where the system's complexity can increase significantly, and tracking actions can be challenging without centralized logging.
- **Security analysis and incident response**: In the event of a security incident, audit logs serve as a critical source of information for analyzing the attack, understanding the attack vectors, and formulating the response. They offer an invaluable, chronological record of events leading up to the incident, allowing security teams to identify the root cause and take appropriate remediation steps.
- **Compliance and regulatory requirements**: Many industries are subject to stringent regulatory requirements that mandate the logging of certain types of activities, particularly those involving access to sensitive data. Audit logging policies can ensure that your service mesh is logging the necessary data to meet these compliance requirements, regardless of whether your services are deployed in a single cluster, multiple regions, or a multi-cloud environment.
- **Anomaly detection**: In addition to tracking regular operations, audit logs can also be used to detect anomalies or irregular activities that might suggest a security breach. For example, repeated failed attempts to access a service could be indicative of a brute force attack.

To summarize, audit logging policies in a control plane help create a clear audit trail of activities within the service mesh. They provide the mechanism for accountability, security analysis, regulatory compliance, and anomaly detection, making them an integral part of the security strategy for any service mesh, irrespective of the deployment being in a single cluster, multi-region, or multi-cloud Kubernetes environment.

Network policies

Network policies in a service mesh control plane are instrumental in managing network traffic at a granular level, controlling who can talk to whom in the network. This fine-grained control is crucial for maintaining the security of the service mesh.

Here are some of the challenges that network policies address in a control plane:

- **Isolation and segmentation**: In a large and complex service mesh, isolation and segmentation are critical to limit the scope of potential security risks. Network policies enable you to define which services can interact with each other, creating isolated segments within the mesh. This segmentation is particularly vital in multi-cluster and multi-cloud Kubernetes environments, where you might have services with different trust levels operating across multiple regions or cloud providers.

- **Traffic control**: Network policies can control both ingress and egress traffic, restricting the flow of data to and from different parts of your service mesh. These restrictions can help prevent unauthorized access to sensitive services and protect against data exfiltration.

- **Risk mitigation**: By limiting the communication between services, network policies can significantly reduce the risk of lateral movement in case of a security breach. If a malicious actor manages to compromise one service, they are confined to that segment and cannot easily spread the attack to other parts of the mesh.

- **Compliance**: Certain compliance regulations require businesses to implement strict network controls to protect sensitive data. Network policies can help satisfy these requirements by providing fine-grained control over network traffic.

In a nutshell, network policies in a control plane provide a robust mechanism for managing network traffic within a service mesh. They help implement isolation and segmentation, control ingress and egress traffic, mitigate security risks, and achieve compliance goals. Whether your service mesh is deployed in a single cluster, multiple regions, or across multiple cloud environments, network policies play a crucial role in maintaining its security.

Peer authentication policies

Peer authentication policies in a service mesh control plane establish a vital layer of security by enabling mutual authentication, typically via mTLS, between microservices in the mesh. Here is a look at some key problems that peer authentication policies address:

- **Validation of service identities**: In a complex service mesh, it is crucial to ensure that services communicating with each other are indeed who they claim to be. Peer Authentication Policies require services to present valid certificates, providing a reliable means of identity validation. This becomes more critical as you scale up to multi-region and multi-cloud environments, where services may span across numerous clusters and cloud providers.

- **Protection against eavesdropping and man-in-the-middle attacks**: Without encryption, data transmitted between services is vulnerable to eavesdropping or interception by malicious actors. Peer authentication policies enforce mTLS, which provides encryption at the transport layer. This effectively protects against such attacks, securing data in transit regardless of whether the environment is a single cluster, multiple regions, or spans multiple cloud providers.
- **Non-repudiation**: With mTLS, not only is the connection secure, but the source of any communication can also be verified. This assurance of non-repudiation can be essential in scenarios where tracking the origin of a request is important for auditing or regulatory purposes.
- **Trust establishment**: In multi-cloud environments, services could reside on different clouds managed by separate teams. Peer Authentication Policies can help establish trust among these services by enforcing strict identity verification.

To summarize, peer authentication policies in a service mesh control plane play a significant role in ensuring that service-to-service communication remains secure. They solve critical security problems by ensuring service identity validation, protecting against certain types of attacks, providing non-repudiation, and establishing trust among services. Whether you are operating in a single cluster, multi-region, or multi-cloud Kubernetes environment, peer authentication policies are instrumental in preserving the integrity of your service mesh's security.

Applying these security policies in the control plane is essential for maintaining a secure service mesh. This is true whether you are operating in a single cluster, multi-region, or multi-cloud Kubernetes environment.

Threat modelling and security best practices

Threat modelling is a proactive approach to identifying, understanding, and mitigating potential threats to a system, and it has become an integral part of the secure design process. In the context of a service mesh, threat modelling helps to anticipate potential security vulnerabilities and to establish best practices to mitigate them.

In the threat modelling process, it is crucial to consider the various components of your service mesh - services, control planes, data planes - and how they interact in different scenarios, from single-cluster to multi-region and multi-cloud environments.

We will not go into too much depth here, as you could author a whole book on threat modeling. However, here are some potential topics to discuss under this section:

- **Identify threats**: Discuss how to enumerate potential threats to the service mesh. This can include insider threats, system vulnerabilities, network breaches, or even seemingly benign misconfigurations.

- **Model threats**: Detail how to model these threats based on their potential impact and exploitability. This could involve creating diagrams to visualize attack vectors or establishing scoring methods for ranking threats.
- **Mitigate threats**: Provide concrete mitigation strategies for the identified threats. This may include secure coding practices, network segmentation, rate limiting, implementing mTLS, and so on.
- **Security best practices**: Discuss security best practices in the context of a service mesh. This could encompass things like principle of least privilege, regular security audits, timely patching and updates, secure handling of certificates and keys, and continuous monitoring and alerting.
- **Case studies of threats and mitigation**: Provide real-world examples of threats to service meshes and how these threats were mitigated. By the end of this section, readers should have a comprehensive understanding of the potential threats to a service mesh and best practices for securing it in single-cluster, multi-region, and multi-cloud Kubernetes environments.

Security considerations

As organizations are increasingly deploying applications in multi-cloud environments, the importance of implementing robust security measures that span across these environments cannot be overstated. A service mesh plays a critical role in such a setup by providing a unified layer of security, control, and observability.

Refer to the following figure:

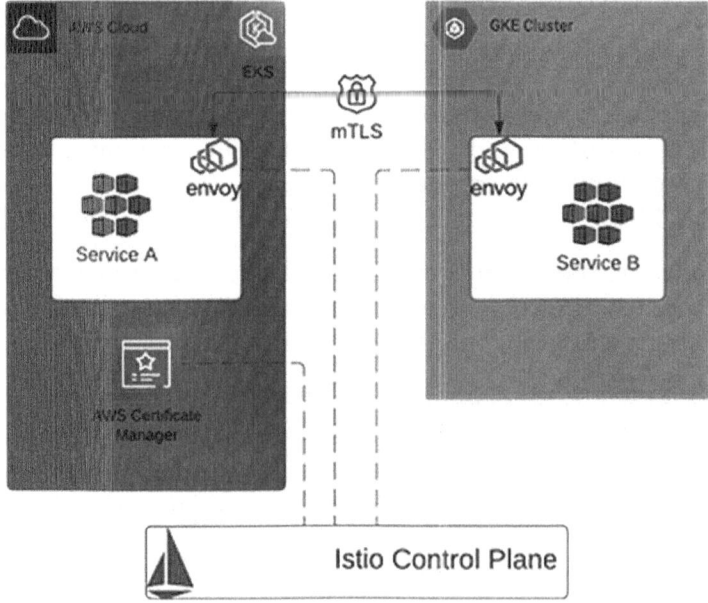

Figure 7.2: Multi-cloud Kubernetes with zero trust

Here are some general, high-level topics to make sure you address while implementing your service mesh security posture:

- **Consistent security policies**: In a multi-cloud environment, it can be challenging to implement consistent security policies across diverse infrastructure. We will explore how to achieve consistency using a service mesh across different cloud providers.
- **Identity and Access Management (IAM)**: Ensuring that only authorized entities have access to your services and data is paramount. We will discuss how service mesh can facilitate centralized IAM across multiple cloud platforms.
- **Data protection**: In multi-cloud environments, data may move between different cloud platforms. We will discuss considerations regarding data encryption in transit and at rest, and how service mesh can ensure secure communication across different cloud environments.
- **Visibility and monitoring**: It is crucial to have a clear understanding of what is happening in your service mesh at all times, especially in a complex multi-cloud environment. We will explore how the observability features of a service mesh can help monitor and maintain security across different cloud platforms.
- **Compliance**: When operating in a multi-cloud environment, organizations need to adhere to various regulations and standards. We will discuss how service mesh can help with ensuring compliance across different cloud environments.
- **Resilience to cloud-specific threats**: In a multi-cloud environment, services could be exposed to different types of threats specific to each cloud provider. We will talk about how to build a resilient service mesh that can withstand these varying threats.

By addressing these considerations, organizations can build a secure and resilient service mesh that spans across their multi-cloud Kubernetes environment.

Conclusion

In this chapter, we delved into the crucial topic of service mesh security. We started off by understanding the evolution and the importance of a comprehensive security posture in service mesh. We discussed essential security principles like zero-trust networks, least privilege, mutual authentication, secure communication, fine-grained access control, policy enforcement, observability, secure ingress/egress, and auditability.

We further explored the concept of traffic encryption using mTLS, and tackled topics such as authorization, access control, and network policies. We also learned about various security policies in a control plane including mTLS policies, authorization policies, audit logging policies, network policies, and peer authentication policies.

Towards the end, we looked at threat modelling, security best practices, and various security considerations to bear in mind when implementing service mesh in multi-cloud Kubernetes environments.

Congratulate yourself for reaching this far! Understanding service mesh security is a critical milestone in mastering the art of managing and securing a service mesh. As we move forward, remember that these security principles and concepts will be integral to our end goal: building a fully enabled, multi-cloud Kubernetes environment powered by Istio and Envoy.

This method is gaining popularity due to its efficiency and reliability in automating deployments and maintaining system consistency. As we step into this new terrain, we will learn how to leverage GitOps for managing our multi-cloud Kubernetes environment with a service mesh, adding another crucial skill to our DevOps arsenal.

Join our book's Discord space

Join the book's Discord Workspace for Latest updates, Offers, Tech happenings around the world, New Release and Sessions with the Authors:

https://discord.bpbonline.com

CHAPTER 8
GitOps Method of Workload Deployment

Introduction

In this chapter, we will take a deep dive into the innovative and increasingly popular method of workload deployment known as GitOps. Originating from the world of **development and operations (DevOps)**, GitOps leverages the power of Git, a widely-used version control system, as the single source of truth for both system and application configuration. This approach takes the concept of **Infrastructure as Code (IaC)** to another level, allowing for faster and safer deployment cycles.

Structure

Here is an overview of the topics we will cover:
- Introduction to GitOps
- Core principles of GitOps
- Benefits and challenges of GitOps
- GitOps workflow
- GitOps tools
- Implementing GitOps in a service mesh

- Securing GitOps workflow
- GitOps best practices

Objectives

By the end of this chapter, you will have gained a thorough understanding of the concept and principles of GitOps. You will be able to recognize the benefits and potential challenges of implementing a GitOps workflow and familiarize yourself with the various tools that support GitOps. In the context of a service mesh, you will grasp how GitOps can be implemented to enhance efficiency and reliability. From a security standpoint, you will learn about the considerations involved in a GitOps workflow. Furthermore, you will understand the implications and specific considerations of GitOps in a multi-cloud Kubernetes environment. Lastly, you will be equipped with the best practices for effectively implementing GitOps. Our journey through this chapter will set you on a path to revolutionizing the way you deploy workloads in a service mesh environment. Let us get started!

Introduction to GitOps

GitOps is a novel approach to continuous delivery in cloud-native applications. As an operating model, GitOps leverages Git as the single source of truth for both infrastructure and application code, allowing development and operations teams to use the same tools for their daily tasks. The main idea behind GitOps is to apply the principles and practices that software developers are familiar with, such as version control and pull requests, to infrastructure management. Refer to the following figure:

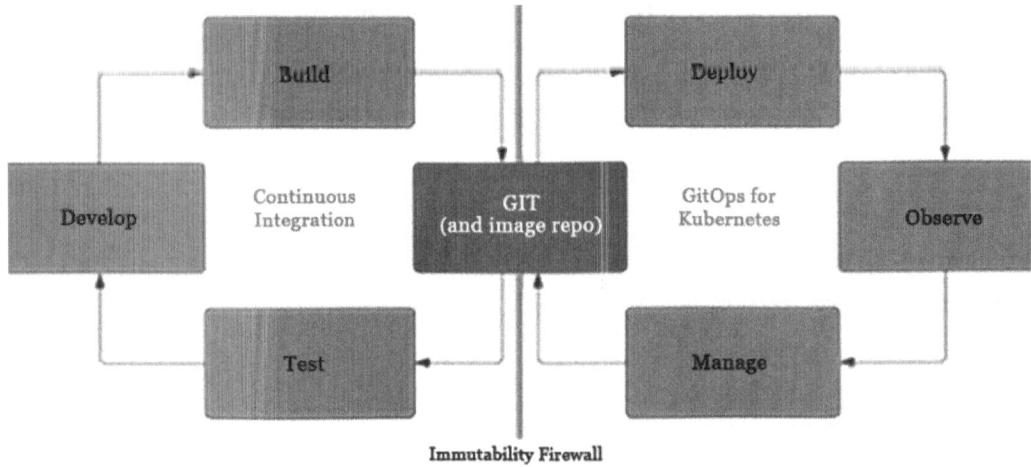

Figure 8.1: GitOps Operating Model

GitOps was coined by *Weaveworks* in 2017, and it is rooted in the core principles of **IaC** and the widespread adoption of Git as the standard version control system for software development. GitOps extends these principles to operational and infrastructure management tasks.

In a GitOps model, changes to infrastructure or application configuration are made in the relevant Git repository and automatically applied to the system. This approach enables a high degree of automation, traceability, and reproducibility since every change is tracked and version-controlled.

The power of GitOps extends beyond simple version control. It also incorporates the benefits of **Continuous Integration and Continuous Deployment (CI/CD)** pipelines, including testing, integration, and deployment. Automated pipelines are triggered by Git events such as push or pull requests, ensuring that only validated and approved changes are promoted to the live environments.

In the context of service mesh and multi-cloud Kubernetes environments, GitOps proves even more potent. It allows for easy management and orchestration of complex workloads across different clusters and clouds, enabling the efficient deployment and seamless rollback of changes.

However, as with any technology or process, GitOps is not the one solution to rule them all. It requires a thorough understanding of its principles, a careful assessment of its benefits and limitations, and a well-thought-out implementation strategy. In this chapter, we will delve into these aspects to provide you with a comprehensive overview of GitOps and its applicability in a service mesh context with multi-cloud Kubernetes.

Core principles of GitOps

GitOps is built on a set of key principles which are crucial in understanding how it operates. These principles play a significant role in ensuring consistency, reliability, and security in multi-cloud Kubernetes environments running a service mesh. Here are the core principles of GitOps.

Declarative infrastructure in GitOps

Declarative infrastructure is one of the core principles of GitOps. This principle focuses on defining and describing the infrastructure setup, application configurations, service mesh configurations, network policies, and all other aspects of the system state in a declarative manner.

The idea is to describe the desired state of the system, and then use automation to achieve and maintain this state. This contrasts with an imperative approach where specific commands or scripts are used to change the system state.

In the context of a service mesh in a multi-cloud, multi-region Kubernetes environment, the declarative approach has several key benefits:

- **Consistency**: By defining your desired state declaratively, you ensure consistency across your entire system. This is important in a multi-cluster and multi-cloud environment, where you need to replicate configurations across different environments or regions. With a declarative approach, the same configuration can be applied consistently across all environments, eliminating discrepancies and reducing the chances of errors.
- **Scalability**: Declarative infrastructure is highly scalable. It enables you to manage large numbers of resources across different clusters and clouds, without having to manually configure each one. You simply define your desired state, and your automation tools ensure that this state is achieved, no matter how many resources you are managing.
- **Version control and auditability**: With the desired state stored as code in a version control system like Git, every change made is versioned. This means you have a complete audit trail of what changes were made, when they were made, and by whom. This enhances accountability and traceability, especially in regulated industries.
- **Reproducibility**: Since your infrastructure is defined declaratively and stored in version control, it is easy to recreate your environment if necessary. This can be extremely useful in disaster recovery scenarios, or when setting up new environments.
- **Integration with service mesh**: Service meshes like Istio and Envoy also operate on a declarative model, the same way a majority of other tools such as ArgoCD, Tekton or Prometheus operate, where the desired state of network traffic, security policies, and observability is defined in a similar manner. This makes GitOps a natural fit for managing service mesh configurations across different clusters and clouds.

In conclusion, declarative infrastructure is a critical principle of GitOps that addresses several key challenges in managing multi-cluster, multi-cloud Kubernetes environments running a service mesh. It ensures consistency, improves scalability, provides a clear audit trail, facilitates reproducibility, and integrates well with service mesh configurations.

Version control as single source of truth

The principle of using version control as the *Single Source of Truth* is a key part of GitOps and has several advantages, especially in the context of managing a complex, multi-cluster, multi-region, and multi-cloud Kubernetes environment, running a service mesh. Some advantages are as follows:

- **Consistency and standardization**: When the version control system (like Git) is treated as the single source of truth, it ensures that all cluster configurations,

service mesh settings, and application deployments are consistent across different environments. This standardization prevents configuration drift and enhances operational predictability.
- **Auditability and accountability**: All changes are tracked in the version control system, providing a detailed audit trail of who made what changes and when. This promotes accountability and can greatly assist in debugging issues and analyzing system changes over time.
- **Reproducibility**: Since the entire system state is captured in version control, it is possible to recreate an identical environment at any point in time. This can be invaluable for testing, disaster recovery, and migration scenarios.
- **Collaboration and transparency**: With a version control system serving as the single source of truth, developers, operators, and other stakeholders can collaborate more effectively. Everyone has visibility into the system state, which promotes transparency and communication across teams.
- **Automated synchronization and rollbacks**: The GitOps principle entails automated synchronization of the system state with the state described in the version control system. If there is a divergence, the system automatically reverts or applies changes to match the desired state. This reduces the risk of manual errors and facilitates rollbacks to a well-known state if an issue arises.

In summary, using version control as the single source of truth within a service mesh context in multi-cloud Kubernetes environment, streamlines management, enhances security, and fosters better team collaboration. This ensures that despite the complexity of the environment, all services remain consistent, traceable, and easy to manage.

Automated synchronization

Automated synchronization is one of the foundational principles of GitOps. It emphasizes automatic and continuous alignment of the system's actual state with the desired state declared in a version control system. This principle offers several benefits, especially when managing multi-cloud Kubernetes environments running a service mesh:
- **Efficiency**: It allows developers and operations teams to focus on defining what the system should look like, not how to achieve that state. This reduces the time and effort spent on manual synchronization tasks and enables teams to be more productive.
- **Consistency**: Automated synchronization ensures that all Kubernetes clusters, across different regions and clouds, are consistently configured as per the specifications defined in version control. This is particularly useful for maintaining consistency in service mesh configurations across various clusters.
- **Resilience**: If the actual state of the system deviates from the desired state due to failures or errors, the system automatically corrects itself, reducing downtime and

ensuring high availability. This is vital in a service mesh setup where maintaining connectivity between services is crucial.

- **Scalability**: As the number of clusters and services increases in a multi-cloud setup, manual synchronization becomes unfeasible. Automated synchronization allows the infrastructure to scale effectively without a corresponding increase in operational complexity.
- **Faster recovery**: In case of a configuration error or a system failure, automated synchronization allows quick rollbacks to a previous desired state. The history of all desired states is maintained in version control, making it easy to revert to a known good configuration.
- **Reduced errors**: Automated synchronization minimizes the risk of human error that can occur during manual interventions. This is especially crucial in a multi-cluster, multi-region, and multi-cloud Kubernetes environment where errors can have significant impacts.

In conclusion, automated synchronization in GitOps offers a reliable and efficient method for managing complex, multi-cloud Kubernetes environments with a service mesh. It provides a mechanism to ensure system consistency, resilience, and scalability while minimizing operational errors and facilitating faster recovery when issues arise.

Immutable infrastructure and image-based deployments

Immutable infrastructure and image-based deployments are key components of the GitOps model. In this setup, environments are never modified after they are deployed. Instead, changes are made in version control and new instances of the environment are deployed. Container images are used for these deployments, which are built once and can be promoted across different environments without changes.

The following figure depicts immutable infrastructure:

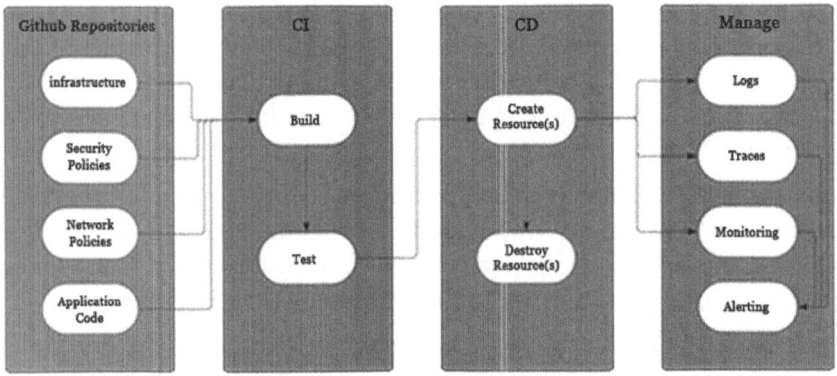

Figure 8.2: Immutable infrastructure diagram

Let us look at the problems this approach solves in the context of a multi-cloud Kubernetes environment running a service mesh:

- **Configuration drift**: One of the biggest problems in managing multi-cluster and multi-cloud Kubernetes environments is configuration drift, where the actual state of the environment deviates from the desired state over time. Immutable infrastructure addresses this by ensuring that each deployment is consistent and replicable, as it uses the same container images across all environments. If a change is needed, a new version of the container image is created and deployed.
- **Deployment consistency**: In a multi-cloud environment, each cloud provider may have different ways of managing services and resources. By using container images for deployments, you can ensure consistency in the application environment across all cloud providers.
- **Reduced failures**: Traditional methods of deploying updates, such as patching running services, can lead to failures due to differences in environment configurations. With image-based deployments, all dependencies are packaged with the application, reducing the risk of failures due to environmental differences.
- **Rollbacks**: In case of an issue, GitOps practices make it easy to roll back to a previous state by deploying a previous version of the container image. This can be crucial in maintaining uptime in a service mesh where different services are dependent on each other.
- **Security and compliance**: Immutable infrastructure reduces the attack surface as there is less chance of configuration tampering in the deployed environment. Moreover, since all changes are tracked in version control, it becomes easier to audit changes and maintain compliance.

Overall, the principles of immutable infrastructure and image-based deployments in GitOps offer a robust method for managing deployments in a multi-cloud Kubernetes environment with a service mesh. This approach provides consistency, reliability, and security while also simplifying the process of rolling back changes when necessary.

Operational procedures through Pull Requests

The principle of operational procedures through Pull Requests is an important part of GitOps that essentially democratizes and streamlines the deployment process in the context of a service mesh running across multiple Kubernetes clusters and cloud environments.

Let us explore the problems it solves:

- **Standardization and consistency**: GitOps encourages the use of Git workflows for operational procedures. This means that changes are made via pull requests, which promotes standardized processes across the development and operations teams. This can be especially helpful in a multi-cloud Kubernetes environment, where uniformity in deployments is crucial to maintaining consistency across different clusters and regions.

- **Peer review**: With the pull request model, every change to the environment is visible to all team members and can be reviewed and approved before being merged. This collaborative approach ensures that errors can be detected early and that knowledge about changes is shared across the team.
- **Traceability and auditability**: All changes made through pull requests are recorded in the Git history, providing a clear audit trail. This level of traceability is important for both troubleshooting and compliance purposes, especially in complex multi-cloud environments.
- **Improved security**: Changes through pull requests allow for more controlled and monitored changes, minimizing the risk of unauthorized or malicious alterations. In the context of a service mesh, where secure communication is paramount, this process contributes to the overall security posture.
- **Synchronization and concurrency**: In a multi-cluster or multi-region Kubernetes setup, synchronization of changes can be challenging. Implementing changes through pull requests allows these changes to be rolled out concurrently and in a controlled manner across all environments.
- **Reduced risk of human error**: Automating operational procedures via pull requests helps to minimize the risk of human error. With automated tests and approval workflows, potential issues can be caught before they impact the live system.

In summary, operational procedures through pull requests offer a method of change management that is consistent, secure, and auditable. It encourages collaboration, improves security, and reduces the likelihood of error in the deployment process - all crucial elements when managing a service mesh across multi-cloud Kubernetes environments.

Wrapping up GitOps principals

These principles come together to form a solid and scalable foundation for managing complex, multi-cloud Kubernetes environments running a service mesh. The declarative nature of GitOps aligns well with Kubernetes and service mesh configurations, enabling teams to handle large-scale, distributed systems efficiently. It allows for precise control and observability of the system state, which is crucial in a multi-cloud setting where workloads are spread across various clusters and regions.

Benefits and challenges of GitOps

GitOps provides a range of benefits but also presents some challenges, especially when managing a multi-cloud Kubernetes environment using a service mesh to connect applications across different clusters.

Benefits of GitOps

Let us now go over the benefits of GitOps:

- **Improved deployment speed and consistency**: GitOps simplifies and automates the deployment process, making it faster and more consistent. The declarative configuration, automated workflows and continuous delivery allow for increases in speed while having a single source of truth, and version control allow for greater consistency. This is particularly beneficial in a multi-cloud environment where consistency across multiple platforms is crucial. In addition, cloud-agnostic, reduced vendor lock-in and centralized management make multi-cloud deployments that much easier to manage.
- **Better auditability**: GitOps leverages the version control system as the single source of truth, providing a complete history of changes, who made them, and when they were made. This transparency is valuable for auditing and compliance, especially in multi-cloud setups where tracking changes can be complex.
- **Enhanced security and compliance**: Changes made through Git are transparent and using pre-commit hooks or pipeline steps can be automatically checked against security and compliance rules. This helps minimize security risks and maintain compliance in a service mesh, where secure communication is paramount.
- **Increased developer productivity**: Developers can use the same Git workflows they are familiar with for operations tasks, increasing productivity. In a service mesh environment, this allows developers to focus on building and improving applications, rather than dealing with the intricacies of deployment across multiple clusters and clouds.
- **Better disaster recovery**: GitOps provides an automatic and efficient way to restore system state in case of a disaster, which is vital in a multi-cloud Kubernetes environment where manual recovery would be time-consuming and prone to errors.

Challenges of GitOps

Let us now go over the challenges of GitOps:

- **Learning curve**: Teams may need time to learn and adapt to the GitOps model, especially if they are used to more traditional operations processes.
- **Complexity of managing secrets**: GitOps does not inherently provide a solution for managing sensitive data, such as passwords or API keys. While there are ways to handle this, it adds complexity and requires careful consideration to ensure secrets are not exposed.
- **Risk of over-reliance on automation**: While automation is the strength of GitOps, over-reliance on it can be risky. There may be situations where manual intervention is necessary, and teams should be prepared to handle such scenarios.

- **Need for robust testing and verification mechanisms**: The automated nature of GitOps means that changes can be deployed rapidly to production. This requires robust automated testing and verification mechanisms to ensure that erroneous changes do not disrupt the system.
- **Managing multi-cluster synchronization**: In a multi-cluster, multi-cloud Kubernetes environment, ensuring synchronization and consistency across all clusters can be a challenge.

In conclusion, GitOps brings numerous benefits to managing a service mesh in a multi-cloud Kubernetes environment, from improving deployment speed to enhancing security. However, it also presents challenges that teams should be aware of and plan for. With the right practices and tools, these challenges can be effectively managed, making GitOps a powerful tool for managing multi-cloud Kubernetes environments.

GitOps workflow

The GitOps workflow is essentially an operational model that uses Git as the system of record for declarative infrastructure and applications. This approach greatly simplifies the process of deploying and managing applications across different environments, including multi-cloud Kubernetes and service meshes.

Let us explore each part of the workflow:

- **Infrastructure as Code**: In a GitOps workflow, you store all system configurations, from infrastructure settings to application definitions, to service mesh configurations, in a Git repository as code. This makes it possible to manage all these components in the same way you manage your application code, using the same version control systems, the same review processes, and the same automation tools. This is particularly powerful in a multi-cloud Kubernetes environment as it allows you to manage the configuration of multiple clusters and even multiple clouds in a standardized way.
- **Version control**: One of the strengths of GitOps is that it uses Git as a version control system. Every change made to the system, every configuration tweak, every deployment is tracked. This means you have a complete, version-controlled history of your entire system. You can see who made what change, when they made it, and why they made it (through commit messages). This is extremely valuable for debugging and auditing, especially in complex environments like multi-cloud Kubernetes.
- **Pull requests**: In a GitOps workflow, changes to the system are proposed using pull requests. This means that no change is made without peer review, and every change is logged and can be linked back to a specific pull request. This enables a high level of transparency and accountability, as well as the possibility for automated testing and validation of proposed changes before they are applied.

- **Automated deployment**: With GitOps, an operator in the cluster (such as ArgoCD, Flux, and so on) automatically synchronizes the system state to match the declared state in Git once changes are approved and merged into the main branch. This automatic synchronization simplifies deployment and management processes, reduces the potential for human error, and speeds up deployment times. It also allows for easy rollbacks to the previous system state if something goes wrong.

In a multi-cloud Kubernetes environment with a service mesh, the GitOps workflow can help ensure consistency across all clusters, enhance security with clear, version-controlled policy management, and improve team productivity by automating much of the deployment and management process. With its combination of automation, transparency, and accountability, GitOps is a powerful approach to managing complex, multi-cloud Kubernetes environments.

GitOps tools

In the GitOps workflow, a range of tools are employed, each playing a specific role, to deliver a seamless, automated, and reliable system. Here are some key tools that are commonly used in a GitOps workflow:

- **Git**: This is the foundational tool for GitOps. Git is a distributed version control system where all the configuration code is stored. It provides the history of changes, allows for peer review of code changes, and acts as the single source of truth. For more information visit **https://github.com/**.
- **Kubernetes**: Kubernetes is a container orchestration platform that automates deployment, scaling, and management of applications. The desired state of Kubernetes resources is declared in manifest files, which can be stored in a Git repository. For more information visit **https://kubernetes.io/**.
- **Helm**: Helm is a package manager for Kubernetes. It allows developers to create, version, share, and publish Kubernetes resources as charts, which are collections of files that describe a related set of Kubernetes resources. For more information visit **https://helm.sh/**.
- **Flux and ArgoCD**: These are two popular GitOps operators for Kubernetes. They monitor the Git repository and ensure that the state of the Kubernetes cluster matches the state declared in the repository. If any divergence is found, these operators will update the cluster to match the declared state in the Git repository. For more information visit **https://fluxcd.io/**, and **https://argo-cd.readthedocs.io/en/stable/**.
- **Terraform**: Terraform is an open-source tool that enables infrastructure as code. With Terraform, you can create, change, and version your infrastructure safely and efficiently. It can be used to manage a wide variety of service providers as well as custom in-house solutions. For more information visit **https://www.terraform.io/**.

- **Prometheus and Grafana**: These tools are used for monitoring and observability. Prometheus collects and stores metrics, while Grafana allows you to visualize those metrics. For more information visit **https://prometheus.io/**, and **https://grafana.com/**.
- **Jenkins/GitHub Actions/GitLab CI**: These are CI/CD tools that can be used in a GitOps workflow. They automate the building, testing, and deployment of applications.

These tools solve various problems in the GitOps workflow. Git, as a version control system, allows tracking changes and maintaining a history of versions. Kubernetes enables managing and scaling applications, Helm simplifies deployment on Kubernetes, and Flux/ArgoCD continuously synchronize the state of the cluster with the state declared in Git. Terraform allows managing infrastructure as code across different platforms. Sealed Secrets/SOPS ensure that sensitive data is stored and managed securely. Prometheus and Grafana enable observability, and Jenkins/GitHub Actions/GitLab CI automate the entire build-test-deploy process. For more information visit **https://www.jenkins.io/**, **https://docs.github.com/en/actions**, and **https://gitlab.com/**.

In a GitOps workflow, these tools work together to deliver an automated, reliable, and traceable system that greatly simplifies the tasks of deploying and managing applications, particularly in complex environments such as multi-cloud Kubernetes running a service mesh.

Implementing GitOps in a service mesh

Implementing GitOps in a service mesh such as one using Istio and Envoy involves leveraging GitOps principles to manage, deploy, and synchronize the configuration of the service mesh. Here is how it can be done across different environments.

Single cluster

In a single cluster, GitOps can be applied to manage Istio or Envoy configurations in a version-controlled manner. The desired state of the service mesh, including routing rules, retry policies, circuit breaking settings, and security policies, is defined declaratively in configuration files stored in a Git repository.

You can use GitOps tools like Flux or ArgoCD to watch for changes in the Git repository. When a change is committed, these tools automatically apply the change to the Kubernetes cluster, thus ensuring that the actual state of the cluster matches the desired state defined in Git.

Multi-region clusters

In a multi-region setup, you might have multiple Kubernetes clusters across different regions, but all clusters could be part of the same service mesh controlled by a global control plane.

In this case, the GitOps workflow can still apply. Each cluster may have its own set of configuration files in the Git repository, or you could structure your configuration files to apply to all clusters universally.

GitOps tools will synchronize the configuration changes to each cluster separately, ensuring all regional clusters align with the global service mesh configuration.

Multi-cloud Kubernetes clusters

Implementing GitOps in a multi-cloud environment adds another layer of complexity as you could have clusters running on different cloud platforms. However, GitOps can greatly simplify multi-cloud management by providing a uniform workflow across different platforms.

In this setup, you would have separate sets of configuration files for each cloud provider, all stored in the same Git repository. The GitOps operator running in each cluster watches for changes in the configuration files related to its cloud provider and synchronizes the changes accordingly.

Moreover, by using a service mesh like Istio and Envoy, you ensure consistent policies, traffic control, and security mechanisms across clusters, irrespective of the underlying cloud platform.

Overall, implementing GitOps in a service mesh, whether in a single cluster, multi-region, or multi-cloud setup, enables more reliable deployments, easier rollback, improved auditability, and greater developer productivity. It brings together the development and operations team to streamline and automate tasks, leading to faster, safer, and more frequent updates.

Securing GitOps workflow

Securing your GitOps workflow is essential to ensure the integrity, confidentiality, and availability of your services and applications, especially in a multi-cloud Kubernetes environment. This section will cover various measures to secure the GitOps workflow:

- **Role-Based Access Control (RBAC)**: Implement RBAC in your Git repositories and Kubernetes clusters. This ensures that only authorized individuals can make changes to the repository and the cluster.

- **Code review and approval process**: All changes to the Git repository should go through a rigorous code review process before being merged. This can help spot potential security issues in the changes.
- **Secure Git practices**: Enable features like GPG signing of commits and two-factor authentication on your Git repository. You can also configure branch protection rules to ensure safe merging of branches, for example a minimum number of code reviewers' approval for a Pull Request, or a set up passing tests that much complete before being allowed to merge into protected branches. This ensures that only authenticated and verified changes are accepted.
- **Automated security checks**: Incorporate automated security checks into your GitOps pipeline. These checks could be static code analysis tools, container image scanners, or configuration security audit tools.
- **Encrypt sensitive data**: All sensitive data, such as passwords, API keys, or other secrets, should be encrypted before being stored in the Git repository. Tools such as Sealed Secrets or HashiCorp Vault can be used for this purpose.
- **Monitor and audit**: Implement strong monitoring and auditing practices. The GitOps model inherently provides an audit trail for changes. However, this should be complemented with logs and monitoring tools for the Kubernetes environment to detect and respond to any suspicious activities.
- **Update and patch regularly**: Keep your GitOps tools and the applications they manage up-to-date with the latest patches and versions. This ensures that you have the latest security enhancements and vulnerability fixes.
- **Principle of least privilege**: Apply the principle of least privilege to all levels of the GitOps workflow, from access to the Git repository to deployment permissions in the Kubernetes clusters.

Remember that securing the GitOps workflow is not a one-time activity but an ongoing process that should be continuously improved and adjusted as per changing requirements and emerging threats. It is an essential part of managing a secure and resilient multi-cloud Kubernetes environment using a service mesh.

GitOps best practices

When implementing GitOps practices in an environment leveraging Istio and Envoy service mesh across single cluster, multi-region, and multi-cloud Kubernetes clusters, the following best practices can be extremely beneficial:

- **Use namespaces and RBAC**: Leverage namespaces in your Kubernetes clusters to provide isolation for your services. Pair this with RBAC to provide fine-grained access and operation control within your cluster.

- **Define clear policies**: Ensure you have clearly defined policies for your service mesh and that these policies are enforced in your GitOps workflow. This includes traffic routing, resiliency, and security policies.
- **Version control everything**: Version control all aspects of your service configuration. This includes the configurations for your Kubernetes clusters, Istio, Envoy, and your application deployments. This ensures that your entire environment is reproducible at any point in time and facilitates rollbacks if required.
- **CI/CD**: CI/CD pipelines are the heart of GitOps. Ensure your pipeline is robust, can handle rollbacks, and is capable of alerting the team if something goes wrong.
- **Ensure zero downtime deployments**: Use Kubernetes deployment strategies like blue-green or canary deployments to ensure zero downtime during application updates.
- **Encrypt sensitive data**: Use Kubernetes secrets or third-party solutions to encrypt and manage sensitive information. Never store plain-text sensitive information in your Git repositories.
- **Automated synchronization**: Ensure automated synchronization between your Git repositories and Kubernetes clusters, which is a core principle of GitOps. This helps in maintaining the desired state of the system and reduces manual errors.
- **Frequent, small, and reversible changes**: Make changes to your system frequently, in small increments, and ensure they are reversible. This way, if something goes wrong, you can quickly rollback to a stable state.
- **Monitor and audit**: Regularly monitor your GitOps workflow and Kubernetes environment to catch and fix issues early. The audit trail provided by GitOps can be invaluable for troubleshooting.
- **Training and awareness**: Finally, ensure that your team is well-trained and aware of GitOps principles, Kubernetes, and service mesh concepts. This is critical for the effective operation and troubleshooting of your environment.

By following these best practices, you can implement a robust and efficient GitOps workflow in a multi-cloud Kubernetes environment leveraging Istio and Envoy service mesh.

Conclusion

Congratulations on completing this chapter in our journey into the world of multi-cloud Kubernetes! This chapter took a deep dive into the GitOps method of workload deployment, a principle that is rapidly gaining traction as a standard for managing and deploying complex, distributed systems in a cloud-native world.

We started by introducing the concept of GitOps and its core principles, emphasizing its significance in environments running single cluster, multi-region, and multi-cloud Kubernetes, and employing a service mesh to connect applications across clusters. Through discussions on declarative infrastructure, version control as the single source of truth, automated synchronization, immutable infrastructure, and the adoption of operational procedures through pull requests, we elucidated how GitOps enhances reliability and reproducibility of deployments.

We then navigated through the benefits and challenges of GitOps, while keeping in context the multi-cloud Kubernetes environments. We delved into the GitOps workflow, taking you from high-level overviews to low-level details of how GitOps contributes to effectively managing service meshes and security policies.

Following this, we explored a variety of GitOps tools and demonstrated how they collaborate to enforce GitOps principles, resolve challenges, and streamline your deployments. We further discussed the practical implementation of GitOps within an Istio and Envoy service mesh, across different types of Kubernetes environments. The chapter culminated in a discussion on securing your GitOps workflow and the best practices to follow when applying GitOps in a service mesh environment.

You have made significant progress on this learning journey, and you should be proud of your accomplishment. Your understanding of GitOps will serve as a solid foundation as we advance further.

In the next chapter, we will build upon this knowledge by exploring the GitOps method of policy deployment. As you will discover, the principles of GitOps are not just applicable to application deployments, but also to the management of policy in our complex environments. Exciting times lie ahead, so stay tuned!

Join our book's Discord space

Join the book's Discord Workspace for Latest updates, Offers, Tech happenings around the world, New Release and Sessions with the Authors:

https://discord.bpbonline.com

CHAPTER 9
GitOps Method of Policy Deployment

Introduction

In the previous chapter, we learned how the GitOps methodology revolutionizes workload deployment in a multi-cloud Kubernetes environment. Now, we are going to apply the same principles to policy deployment. This approach ensures your environments are not only consistent and easily reproducible, but also conformant and compliant with the policies set by your organization.

Structure

Here is an overview of the topics we will cover:
- Introduction to GitOps policy deployment
- Policy as code
- Core principles of GitOps for policy deployment
- Implementing GitOps for policy deployment
- Tools for GitOps policy deployment
- Securing GitOps policy deployment
- GitOps policy deployment best practices

Objectives

By the end of this chapter, you will understand the key concepts behind GitOps policy deployment and how it can streamline and secure your multi-cloud Kubernetes environments. You will be familiarized with the tools used in GitOps for network and security policy deployment and understand how to select the ones that best suit your organization's needs. Additionally, you will learn about the best practices to follow when implementing GitOps for policy deployment, so you can avoid common pitfalls and ensure a smooth, efficient deployment process. The knowledge and skills you gain in this chapter will be invaluable in designing and managing your organization's service mesh.

Introduction to GitOps policy deployment

In the age of cloud-native technologies, managing policies across multi-cloud Kubernetes clusters can be quite a challenge, particularly when a service mesh like Istio and Envoy is involved. This is where the concept of GitOps comes in, extending its principles from workload deployment to policy deployment.

As you may remember, GitOps is an operational model that places Git at the heart of defining and controlling the state of infrastructure. It essentially applies the Git distributed version control system and its best practices, such as pull requests and version tracking, to infrastructure management.

When it comes to policy deployment, the GitOps approach offers significant benefits for both consistency and reliability. This model treats policy as code, enabling version control, collaboration, and **Continuous Integration and Continuous Deployment (CI/CD)** for policy changes, just like we do for software development.

In a single Kubernetes cluster, GitOps can ensure that your policies are consistently applied and easily trackable. However, the real power of GitOps policy deployment comes into play in multi-region and multi-cloud Kubernetes environments. By managing policy as code, you can ensure consistency of policy enforcement across all your clusters regardless of their location, whether they are on-premises or in different clouds.

A well-designed service mesh using Istio and Envoy further amplifies these benefits. It helps implement fine-grained, application-level policies across your clusters. With GitOps, these policies can be version controlled, tested, and applied systematically, greatly reducing the risk of configuration errors and security breaches.

In this chapter, we will dive deeper into the concept of GitOps policy deployment, its principles, implementation, and the tools that can help you leverage its full potential. We will also discuss how to secure your GitOps policy deployment and share some best practices for your multi-cloud Kubernetes environments.

Policy as code

The concept of **Policy as Code (PaC)** is a fundamental principle in the GitOps method of policy deployment. It treats system and security policies much like software code, applying the same principles of version control, automated testing, and deployment. Some examples include infrastructure and cloud security, application security and compliance, and operational policies. This approach facilitates collaboration, increases speed of changes, and improves the consistency and accountability of policy enforcement.

When using a service mesh like Istio and Envoy in a single cluster (refer back to the preceding chapter which covers service mesh), PaC allows for precise, automated control over traffic routing, load balancing, and security settings. Policies defined as code are checked into a Git repository, where they can be versioned, peer reviewed, and history-tracked. Any proposed changes to the policy can be validated and tested before being merged and applied, significantly reducing the risk of configuration errors.

When you scale to multi-region and multi-cloud Kubernetes environments, the advantages of PaC become even more evident. Managing consistent policies across multiple clusters and regions manually can be error-prone and difficult to track. By leveraging GitOps and PaC, you can ensure consistent policy enforcement across all clusters regardless of their location. Policy changes can be propagated automatically across all environments, maintaining state consistency.

The Istio and Envoy service mesh plays a vital role in this process. Some examples of this are the traffic routing, and load balancing. A scenario to help illustrate this would be routing 20% of traffic to a new version of a service as result of a network policy. Another example is the enforcement of mTLS as the result of a security policy. By implementing policy enforcement at the application level, it allows for granular control over communication between services across your clusters. With PaC, these policies can be version controlled, systematically applied, and easily audited, enhancing security and reliability in your multi-cloud Kubernetes environments.

In the following sections, we will discuss the key components of PaC, including policy definition, enforcement, monitoring, and reporting, all with a focus on GitOps and service mesh in a multi-cloud context.

Core principles of GitOps for policy deployment

Implementing GitOps for policy deployment shares many principles with the general GitOps methodology but has some specifics that are crucial when it comes to enforcing policies across the infrastructure. Let us examine these core principles in the context of multi-cloud Kubernetes environments utilizing a service mesh like Istio and Envoy.

Policy as code

As mentioned earlier, this principle involves treating policy definitions much like software code. This means leveraging version control systems (like Git) to maintain policy definitions, track changes over time, and allow for easy rollback if needed. More concrete examples using Git include version control and history, collaboration via Pull Request reviews, automation to achieve **CI/CD** and aiding in auditability and compliance. It allows for greater transparency, accountability, and efficiency.

Automated policy enforcement

With the help of GitOps, policy enforcement becomes an automated process. Once a policy is committed into the repository, the GitOps operator automatically applies it across the environments. The existence of a Git repository is not enough to enable automated policy enforcement, however. The policies must be defined, and mechanisms must be put in place to bridge the gap between the code and the service mesh. There must also be policy validation and testing, monitoring and rollback mechanisms must also be put in place. Various tools such as ArgoCD and GitHub Actions have solutions for these tasks. In a service mesh, this could involve traffic routing rules, access controls, or security settings.

Continuous monitoring and reconciliation

Just as with state of applications in GitOps, the state of policies is continuously monitored, and any divergence from the desired state stored in the repository is automatically corrected. Using a policy agent (like Open Policy Agent Gatekeeper: **https://github.com/open-policy-agent/gatekeeper**) which resides in your clusters and continuously evaluates policies against the state of your service mesh configuration is very effective against preventing policy or configuration creep. This ensures consistency across all environments.

Automated validation and testing

In a similar vein to CI/CD pipelines for code, policies can also be automatically validated and tested before being deployed. A common use case here is when an Engineer commits a new code and opens a PR, another Engineer reviews the change, and it ultimately is merged kicking off the CI/CD pipeline. The automated validation for configuration and syntax runs as a pipeline step and can halt the pipeline if a policy rule is violated so the committing Engineer can review and remediate. Once complete with no violations the code change can be deployed via the normal processes. This can help catch potential issues early, before they impact the live environment.

Observability and auditability

All changes to policies are tracked in the version control system, providing a clear audit trail. Moreover, integration with logging and monitoring systems can provide real-time insights into policy enforcement.

Security and compliance

Since policies are explicitly defined and controlled, it becomes easier to ensure security and compliance requirements are consistently met across all environments. This concept can be better illustrated using an example where we establish our static analysis and security scanning of code as a pipeline step. The following points go into more detail:

- **Policy language features**: Many policy languages like **Open Policy Agent (OPA)** Rego offer built-in capabilities for static analysis. These features can scan policies for potential vulnerabilities, suspicious constructs, or logical fallacies before deployment.
- **Dedicated tools**: Integrate tools like rego-validate or the Styra OPA Gatekeeper Static Analyzer into your development pipeline. These tools provide deeper analysis, identifying potential security risks or compliance violations within your policies themselves.
- **Pipeline integration**: Embed these static analysis steps within your CI/CD pipeline. This ensures every policy change undergoes security checks before reaching production, preventing the deployment of potentially risky configurations.

In the context of multi-cloud Kubernetes and service mesh, these principles help provide a consistent, secure, and efficient approach to managing and enforcing policies, regardless of the scale or complexity of your environment.

Implementing GitOps for policy deployment

The implementation of deploying policies in a Kubernetes environment using a service mesh such as Istio and Envoy is similar to the other deployment setup we have described previously. We can enumerate them here as to how they can be implemented specifically for policies:

- **Centralized policy management**: Utilize a centralized Git repository for managing all policy definitions. This would serve as the single source of truth for all policies applied across your clusters. This includes service mesh configuration policies, network policies, security policies, and so on. The centralized approach ensures consistency and transparency across all environments, whether it is a single cluster, multi-region clusters, or a multi-cloud setup.

- **Version control for policy changes**: Changes to policies should be tracked using version control systems, such as Git. This allows for easy rollback, change tracking, and auditing. Each commit should have a clear and meaningful message to provide context for the change.
- **Automated policy deployment**: Use a GitOps operator or a tool, like ArgoCD, Flux, or Jenkins X, to monitor the Git repository and automatically apply the changes in policy across all clusters when a new commit is pushed. The operator should be configured to continuously monitor and reconcile the state of the clusters with the desired state defined in the Git repository.
- **Policy validation and testing**: Just as you would validate and test code before deployment, it is important to do the same for policy changes. This could be done by integrating policy validation tools into your CI/CD pipeline. For instance, tools like **OPA** can be used for policy validation.
- **Secure access control**: It is important to limit who can make changes to policies. Use access controls provided by Git and your GitOps tools to ensure only authorized users can push policy changes. Additionally, implement peer review practices, such as requiring pull requests and approvals before changes can be merged.
- **Observability and auditing**: Ensure your logging and monitoring solutions provide visibility into policy enforcement and can alert you to any issues. Your version control system should provide a clear audit trail for policy changes, and your GitOps tools should log all actions they take.
- **Consistency across environments**: Regardless of the scale or complexity of your setup, it is crucial to ensure policy consistency across all environments. This is especially important in multi-cloud or multi-region clusters where environmental differences can often lead to unexpected behaviors. GitOps helps ensure this consistency by automatically synchronizing policy state across all environments based on the single source of truth in the Git repository.

Remember, while the above practices provide a general approach to implementing GitOps for policy deployment, you may need to adjust them based on your specific needs, environments, and organizational practices. Always consider the security implications of any changes you make to your workflow.

Tools for GitOps policy deployment

There are a variety of tools available to help implement GitOps principles for policy deployment in a Kubernetes environment using a service mesh such as Istio and Envoy. Here, we will focus on a few widely used ones. Keep in mind many of these tools have features that overlap and would be redundant if both were implemented. Always choose your tooling based off of stakeholder requirements and business use cases:

- **FluxCD**: FluxCD is a toolset for implementing GitOps in a Kubernetes environment. It allows you to use a Git repository as the source of truth for your desired state.

It continuously monitors the repository and applies any changes to your clusters. For policy deployment, FluxCD can ensure policies are applied consistently across your clusters, whether they are single cluster, multi-region clusters, or a multi-cloud setup. Find more on FluxCD here: **https://fluxcd.io/**.

- **ArgoCD**: ArgoCD is a declarative, GitOps continuous delivery tool for Kubernetes. It follows the same GitOps principles as FluxCD, allowing you to use a Git repository as your source of truth for the desired state. It can be used to automate the deployment and synchronization of your policies across different Kubernetes clusters. Find more on ArgoCD here: **https://argo-cd.readthedocs.io/en/stable/**.
- **OPA**: OPA is a general-purpose policy engine that unifies policy enforcement across the stack. It uses a high-level declarative language, allowing you to author policy as code. OPA can be integrated with your CI/CD pipeline to validate policies before they are committed to the repository, helping to prevent erroneous or insecure policies from being deployed. Find more on OPA here: (**https://www.openpolicyagent.org/**).
- **Helm**: Helm is a package manager for Kubernetes that simplifies the deployment of applications and services. In the context of policy as code, Helm charts can be used to package policy definitions and configurations, allowing them to be versioned, shared, and deployed in a standardized manner. Find more Helm here: **https://helm.sh/**.
- **Terraform**: While not specific to Kubernetes or service meshes, Terraform is a powerful tool for provisioning and managing infrastructure as code. It supports a wide range of providers, making it particularly useful in multi-cloud environments. Find more on Terraform here: **https://terraform.io/**.

These tools, when utilized together, can provide a robust GitOps workflow for policy deployment across single cluster, multi-region clusters, and multi-cloud Kubernetes environments using service meshes like Istio and Envoy. Remember, the choice of tools may depend on the specific needs of your project or organization, and these should be carefully evaluated against your requirements.

Securing GitOps policy deployment

Securing the GitOps policy deployment process is vital in any Kubernetes environment, whether it is a single cluster, multi-region, or multi-cloud setup, especially when running a service mesh such as Istio and Envoy. Here are some best practices:

- **Secrets and sensitive data**: Secrets and sensitive data should not be stored in any Git repository. They can safely be stored in tools like Hashicorp Vault, or Kubernetes' own secrets management can be used to safely store this information and to make it available to the workload at build time. Also, if secrets need to be loaded into the container image at build time, perhaps in the form of an environmental variable, mechanisms such as GitHub Action secrets and variables can be used.

- **PR reviews**: Make use of Pull Request reviews to have a second pair of eyes on policy changes. This can help catch potential security issues and ensures that changes are validated by more than one person.
- **Automated scanning**: Implement automated testing for your policies to verify their functionality and impact on the system. This can be part of a CI/CD pipeline, where changes to policies trigger a suite of tests.
- **Least privilege principle**: Follow the least privilege principle when defining roles and permissions in your policies. This principle is ubiquitous and should apply to the entire SLDC. This ensures that services and users have the minimum permissions necessary to perform their tasks, reducing the potential impact of a breach.
- **Separation of duties**: Implement separation of duties where possible. This means that no single individual should have complete control over all aspects of policy deployment, thus reducing the risk of accidental or intentional misuse. A common use case is to have the QA team under a VP of Engineering. There exists a conflict of interest when a VP of Engineering needs to push a feature quickly and the team that is responsible for clearing it for deployment. The QA team needs to operate as a separate team with a clear duty to produce high quality and secure features, rather than just pumping out features. Having them exist side by side ensures high quality and secure features at the fastest pace possible.
- **Continuous monitoring**: Continuously monitor your system for any deviations from the desired state and for any indicators of compromise. Tools like Prometheus and Grafana can provide insights into your system's state and alert you to potential issues. Common metrics for infrastructure health include CPU, memory and storage utilization, container and pod health, network traffic and system logs. Metrics for application security include API request volumes, error rates, response times and auth failures. Metrics for user activity and access control include user logins and activities, privileged user access, and login failures. If you are under any compliance frameworks, they will have a host of metrics to follow as part of their controls.

By following these practices, you can enhance the security of your GitOps policy deployment, ensuring that your service mesh and wider system remain secure and compliant in all types of Kubernetes environments.

GitOps policy deployment best practices

Implementing GitOps for policy deployment in Kubernetes environments, whether single cluster, multi-region, or multi-cloud, requires careful thought and adherence to best practices. When employing a service mesh such as Istio and Envoy, the following best practices can prove beneficial.

Clearly define policies

Policies, in the context of GitOps and Kubernetes, provide a structured framework for how resources, applications, and systems should behave in an environment. They can cover a wide range of aspects, from specifying network access rules to determining the number of resources that can be consumed by a particular service. Good policies are clear and concise, granular, measurable and version controlled. Bad policies are vague or ambiguous, over specific, not measurable and not version controlled.

In the realm of single-clareter, multi-region, and multi-cloud Kubernetes deployments, having clearly defined policies is pivotal for several reasons:

- **Consistency**: Clearly defined policies ensure consistency across different deployments. This is especially crucial in multi-region and multi-cloud environments, where you might be dealing with various service providers and their unique configurations. By defining policies, you ensure that all regions and clouds adhere to the same rules, resulting in consistent behavior across all deployments.
- **Automation**: Clear and defined policies can be automatically applied and enforced, reducing the risk of human error. This is a cornerstone of the GitOps methodology, where infrastructure is managed declaratively, and changes are implemented automatically based on the declared state stored in Git repositories.
- **Security**: Security policies define who can access what, how, and when. By defining these clearly, organizations can ensure the right level of access to the right entities, reducing the chances of unauthorized access or security breaches. This is especially important in multi-cloud environments, where different cloud providers may have different default security configurations.
- **Efficiency**: Having clearly defined policies can result in more efficient operations. When everyone knows the rules and how systems should behave, there is less time spent on debugging and more time for innovation. This becomes even more critical in large-scale deployments across multiple clusters, regions, or cloud providers, where manually handling inconsistencies could be resource intensive. Simple and smaller changes tend to deploy faster, are easier to understand and less impactful in case they have negative consequence. Large scale changes tend to take longer, are harder to understand and therefor are harder to troubleshoot.
- **Compliance**: Particularly for industries that have to adhere to strict regulatory standards, clearly defined policies help in demonstrating and maintaining compliance. A policy that ensures all stored data is encrypted, for example, can be critical for meeting data protection regulations. Each framework has its own set of controls, however having good policies in place help keep you within those controls and allow you to use them as evidence of compliance behavior.

By defining policies clearly, teams can manage complex Kubernetes environments more effectively, making the system behavior more predictable and under control. This translates into a more secure, efficient, and reliable system, regardless of the scale or distribution of the Kubernetes clusters.

Automate everything

The GitOps principle of **automate everything** is one that strongly emphasizes replacing manual processes with automated ones, and this is equally vital when it comes to policy deployment.

Implementing this principle in the context of policy deployment within single cluster, multi-region, and multi-cloud Kubernetes environments solves a number of key issues, such as the following:

- **Reduction of human error**: Automation significantly reduces the risk of human error that is often associated with manual processes. In larger teams this helps keep actions consistent as opposed to varying wildly with many team members contributing in different ways. This is particularly important in complex environments such as multi-region and multi-cloud Kubernetes deployments where the scope for errors can exponentially increase due to the size and complexity of operations.
- **Consistency and standardization**: Automated deployment ensures that the same set of policies is applied consistently across all clusters, regardless of their geographic location or the cloud provider they reside in. This ensures standardization across the entire infrastructure and reduces discrepancies.
- **Efficiency and speed**: Automated policy deployment allows for quicker rollouts of policy updates or new policies across all clusters. This allows for quicker reaction times when changes in policy are required due to regulatory changes or security threats.
- **Scalability**: As your infrastructure grows, the task of manually deploying policies becomes more and more unfeasible. Automation is key to scalability as it allows policies to be applied across a growing number of clusters and nodes without a corresponding increase in manual effort.
- **Compliance and auditing**: An automated policy deployment process creates a detailed and accurate audit trail that can be reviewed for compliance purposes. This is particularly useful in regulated industries that need to demonstrate adherence to certain standards or regulations.
- **Resource allocation**: By automating policy deployment, teams free up human resources that would otherwise be spent on manual tasks. This allows personnel to be reallocated to areas where they can add more value, such as strategic planning or innovation.

By automating everything in your policy deployment process, you are ensuring the reliability and consistency of policy enforcement. This makes your systems more secure, compliant, and manageable, regardless of the complexity and scale of your Kubernetes deployments.

Version control

In the context of GitOps policy deployment best practices, utilizing version control solves several problems and adds substantial benefits in single cluster, multi-region, and multi-cloud Kubernetes environments. While these are also similar to previously defined principals, some key components specific to version control are as follows:

- **Traceability and auditability**: Version control systems provide a detailed history of changes, including who made the change, what the change was, and when it was made. This audit trail is crucial for identifying when and how a policy change occurred, which is especially useful for post-incident reviews and compliance audits.
- **Collaboration and conflict resolution:** Multiple teams often manage policies across different regions or cloud environments. Version control allows these teams to work together more effectively, providing mechanisms to merge changes and resolve conflicts in a standardized way.
- **Repeatability and consistency**: Policies stored in version control can be consistently applied across different clusters, regions, and cloud environments. This ensures that all parts of the system are configured according to the same policies, reducing the risk of inconsistencies that can lead to security vulnerabilities or operational issues.
- **Rollback and recovery**: If a policy change leads to unforeseen issues, version control systems make it easy to revert to a previous state. This allows teams to quickly recover from errors and reduces the impact of mistakes or unforeseen issues.
- **Change management and review**: Changes to policies can be reviewed and approved as part of a structured process before they are applied. This helps to ensure that only valid and appropriate changes are made, reducing the risk of errors or security issues.
- **Automated enforcement**: With version control as the source of truth, automation tools can continuously ensure that the live state of your clusters matches the desired state defined by the policies in version control. If discrepancies are detected, they can be automatically corrected, further enhancing security and compliance.

In summary, the practice of version control in GitOps policy deployment contributes to robust policy management, better collaboration, improved security, and stronger compliance, making it a critical best practice for managing policies across complex Kubernetes environments.

Regularly review and update policies

The principle of regularly reviewing and updating policies is a crucial part of GitOps policy deployment best practices. In single cluster, multi-region, and multi-cloud Kubernetes environments, it brings multiple benefits and solves several challenges, such as the following:

- **Keeping up with changes:** The landscape of technology and security is continually evolving. New vulnerabilities are discovered, new regulatory requirements are put in place, and business needs change over time. Regularly reviewing and updating policies ensures that they stay relevant and effective in the face of these changes.

- **Maintaining security posture**: Policies are a key aspect of maintaining the security posture of your Kubernetes environments. By regularly reviewing them, you ensure that they effectively mitigate new risks and continue to enforce your desired security posture.

- **Ensuring compliance**: In many industries, there are strict regulations governing data handling, privacy, and other aspects of IT operations. Policies need to be regularly reviewed and updated to ensure they continue to meet these regulatory requirements, helping to avoid potential penalties and reputational damage.

- **Improving efficiency**: As your organization and its processes evolve, there might be opportunities to make your policies more efficient. Regular reviews can help identify unnecessary redundancies, streamline processes, and reduce the administrative overhead associated with policy enforcement.

- **Reducing error**: Regular reviews and updates can help catch and correct errors in policies before they lead to more significant issues. This can prevent potential security breaches, operational failures, or other negative impacts.

- **Facilitating continuous improvement**: A regular review and update cycle feeds into a continuous improvement process for your GitOps policy deployment. This iterative process can help refine and improve your policies over time, optimizing them for your specific operational context and evolving needs.

In summary, regularly reviewing and updating policies in GitOps policy deployment helps organizations stay agile and responsive to changes, improves security and compliance, and facilitates continuous improvement in operational efficiency.

Testing and validation

Testing and validation are crucial aspects of GitOps policy deployment best practices and serve several important functions across single cluster, multi-region, and multi-cloud Kubernetes environments:

- **Preventing errors**: Before any changes are merged into the main branch, they can be tested in isolation. This includes any alterations to infrastructure configuration, service mesh rules, or security policies. This ensures that changes will not cause unforeseen disruptions or vulnerabilities when applied to the live system.

- **Ensuring consistency**: With GitOps, your Git repository serves as the single source of truth. By using automated tests and validation steps, you can ensure that the reality of your cluster state aligns with the desired state described in your Git repository. This is especially crucial in multi-region and multi-cloud environments, where consistency across various instances and configurations can be challenging to maintain.
- **Policy compliance**: Automated testing and validation can also enforce policy compliance. Whether you are working in a regulated industry or enforcing your internal rules, policies encoded in your Git repository can be automatically tested and validated. This ensures that all deployments are in line with your policies, reducing the risk of non-compliance.
- **Rollbacks**: If issues are detected during the validation phase, GitOps allows for easy rollbacks to previous configurations. Since all changes are version-controlled and stored in Git, you can revert to an earlier, stable state if an update leads to failure or security issues.
- **Continuous feedback**: Automated tests give developers immediate feedback on their changes. This accelerates the development process and improves the overall quality of deployments.
- **Facilitating learning**: Mistakes are an opportunity for learning and growth. By catching errors early and often through automated testing and validation, teams can understand their weaknesses and improve their practices over time.

In summary, testing and validation in GitOps policy deployment allow organizations to prevent errors, ensure consistency, comply with policies, facilitate learning, and maintain a robust and secure system, no matter the size or complexity of their Kubernetes deployments.

Observability

Observability is a critical best practice for GitOps policy deployment across single cluster, multi-region, and multi-cloud Kubernetes environments. It addresses several problems and concerns such as the following:

- **System transparency**: Observability allows you to understand the internal state of your system from the outside. It gives insights into how your policies are affecting the system's behavior in real-time. With proper observability, teams can gain a clear view of the entire system state, reducing the time needed to detect and diagnose potential issues.
- **Efficient troubleshooting**: When a deployment fails or a policy unexpectedly impacts the system, observability tools can help pinpoint the root cause of the

issue. Detailed metrics, logs, and traces can provide invaluable context for troubleshooting, allowing you to find and resolve issues faster.

- **Performance monitoring**: Observability can also track the performance of your applications, clusters, and infrastructure over time. This information can be used to identify patterns and trends, understand resource utilization, and optimize performance in the context of the applied policies.
- **Policy impact analysis**: By integrating observability into your GitOps workflows, you can understand the impact of your policies on the system's performance, security, and stability. This enables you to fine-tune your policies and strategies over time, based on observable data, rather than assumptions or predictions.
- **Compliance and audit**: Observability also aids in maintaining compliance and audit trails. By monitoring all system activities and changes, you can ensure that all operations are adhering to your organization's policies and regulatory standards.
- **Enhancing collaboration:** Observability tools provide a shared source of truth for all stakeholders. This includes developers, operations, security teams, and even business units. This shared perspective can help enhance collaboration and decision-making.

In multi-region and multi-cloud Kubernetes environments, the role of observability becomes even more important due to the increased complexity and potential for inconsistencies between different environments. Observability tools help maintain consistency and ensure that policies are being applied and functioning as expected across all environments.

In conclusion, observability in GitOps policy deployment helps ensure system transparency, efficient troubleshooting, performance monitoring, policy impact analysis, and compliance. It also enhances collaboration, making it an essential best practice in managing complex Kubernetes deployments.

Plan for failure

Even with the most carefully designed systems and robust policies, failures are bound to happen at some point. This is especially true in complex environments like single cluster, multi-region, and multi-cloud Kubernetes deployments. By planning for failure as a part of your GitOps policy deployment best practices, you solve several key problems such as the following:

- **Minimized downtime**: By having a plan in place for potential failures, you can reduce the time it takes to restore service after a failure occurs. This is achieved by having clearly defined procedures for diagnosing and resolving common failure modes, thereby reducing the impact on users and services. For example, let us imagine a scenario where you deploy a new version of your service using a rolling update strategy across your clusters. However, shortly after the update starts, users report issues and performance degradation. Having monitoring in place allows you to detect and diagnose the issue quickly and an automated rollback is initiated quickly and safely using your automated rollback procedure. Your **Mean Time to Recovery (MTTR)** is severely reduced in this scenario compared to if there were no plan for failure in place.

- **Mitigated data loss**: A solid failure plan should include strategies for data backup and recovery. This ensures that critical data can be restored quickly if a failure causes data loss or corruption. Using the example above a rollback procedure which accommodates for data migrations will include the preservation of the correct data set for the version of active code.
- **Ensured business continuity**: Planning for failure includes creating continuity plans that allow your business to continue operating even in the face of system failures. This can involve redundant systems, failover mechanisms, and disaster recovery strategies. In the previous example your automated procedures allow for quick rollback with no downtime or bad user experience, preserving your business continuity goals.
- **Stress reduction**: When failure occurs, having a plan in place can reduce stress and panic. Teams know exactly what steps to follow, reducing the potential for errors caused by rushed decision-making or confusion. Watching your procedures execute with the smooth preservation of data, and uptime allow for stress free deployments.
- **Learning and improvement**: Planning for failure also means post-mortem analysis to understand what went wrong and how it can be prevented in the future. This continuous learning approach leads to system and process improvements that increase resilience over time.

In the context of a multi-region and multi-cloud Kubernetes environment, planning for failure becomes even more critical. These environments are inherently complex and a failure in one region or cloud could potentially have cascading effects on the entire system. Planning for failure in these environments may include strategies for isolating failures to a specific region or cloud, maintaining data consistency across regions and clouds, and ensuring the high availability and resilience of the service mesh and applications running on it.

In conclusion, planning for failure is an important aspect of GitOps policy deployment best practices. It prepares your organization to effectively respond to failures, minimizing their impact and turning them into opportunities for learning and improvement.

By following these best practices, you can ensure your GitOps policy deployment process is robust, secure, efficient, and effective in managing your Istio and Envoy powered service mesh across all types of Kubernetes environments.

Conclusion

Congratulations! You have successfully navigated through the complexities of GitOps for policy deployment. In this chapter, we delved deep into the world of GitOps and how it enhances policy deployment in single cluster, multi-region, and multi-cloud Kubernetes environments.

We began by introducing the concept of GitOps for policy deployment, exploring its significance in managing complex systems and enhancing the workflow efficiency. We then moved on to the notion of **PaC**, highlighting its role as an integral part of the GitOps approach and how it streamlines policy management in distributed systems.

In understanding the core principles of GitOps for policy deployment, we discussed various elements such as version control, automated synchronization, and immutability, amongst others. This set the stage for our discussion on how to implement GitOps for policy deployment, where we walked through practical steps to harness the power of GitOps in managing policies for your service mesh, irrespective of whether it spans single cluster, multi-region, or multi-cloud deployments.

We then focused on the various tools available for GitOps policy deployment, their functionalities, and how they can be utilized in a Kubernetes environment. We also explored the security aspects of the GitOps workflow, emphasizing its criticality in maintaining the integrity of your systems and data.

Lastly, we reviewed the best practices for GitOps policy deployment. Here, we discussed different strategies, including clearly defining policies, automating everything, using PaC, regularly reviewing and updating policies, ensuring robust security measures, thorough testing and validation, maintaining observability, and planning for failure.

Having completed this chapter, you are now equipped with the knowledge to implement, secure, and manage GitOps policy deployment for your service mesh in various Kubernetes environments.

As we progress into the final chapters of this book, get ready to dive into another critical aspect of Kubernetes environments, that is, *Proactive Monitoring of the Clusters*. In the next chapter, we will learn about the importance of monitoring, discuss various monitoring tools, techniques, and best practices in the context of Kubernetes and service mesh. Stay tuned!

CHAPTER 10
Proactive Monitoring of the Clusters

Introduction

This chapter focuses on one of the most crucial aspects that govern the effective functioning of your Kubernetes clusters. Monitoring is a critical component in maintaining the performance and health of the clusters and the services hosted on them. Given the complex and distributed nature of Kubernetes and the service mesh, proactive monitoring becomes significantly more important.

Structure

Here is an overview of the topics we will cover:
- Introduction to proactive monitoring
- Importance and benefits of proactive monitoring
- Challenges in proactive monitoring
- Tools and techniques for proactive monitoring in Kubernetes
- Implementing proactive monitoring
- Understanding and using metrics, logs, traces and alerts for proactive monitoring
- Best practices for proactive monitoring

- Securing your monitoring stack
- Proactive monitoring in service mesh

Objectives

This chapter aims to impart a comprehensive understanding of proactive monitoring in a Kubernetes environment. We will explore various facets of proactive monitoring, including its benefits and why it is important, challenges that you may encounter, and how you can navigate around them. We will delve into the different tools and techniques used for monitoring Kubernetes clusters, with a special emphasis on how these apply to a single cluster, multi-region, and multi-cloud environments. We will also lay out the best practices for proactive monitoring to help you ensure that your clusters remain robust, efficient, and resilient. By the end of this chapter, you should be equipped with a solid foundation of knowledge on proactive monitoring in Kubernetes, enabling you to apply it effectively in your deployments.

Introduction to proactive monitoring

Proactive monitoring is a vital aspect of maintaining a healthy and reliable service, especially in complex environments like multi-cloud Kubernetes clusters running a service mesh. It is all about taking a forward-looking approach to managing and maintaining systems to prevent problems before they impact the service's quality and reliability.

Unlike traditional monitoring, which often focuses on detecting and fixing issues after they occur, proactive monitoring involves continuously analyzing the state of your system to predict potential issues before they cause service degradation or outages. It revolves around gathering and analyzing data, metrics, and logs to understand the system's state and behavior over time.

In the context of Kubernetes, proactive monitoring involves the collection and analysis of metrics from various components like the nodes, pods, and containers. It also includes service mesh components when used, such as Istio and Envoy proxies. This data can provide valuable insights into the performance, health, and reliability of the services running on your clusters.

Furthermore, proactive monitoring is not just about the systems themselves. It also involves monitoring the surrounding environment, including network performance and traffic patterns, security threats, and even system usage trends. This can help you anticipate demand spikes, identify potential security issues, and ensure your system can meet your users' needs.

In this new age of DevOps and **Site Reliability Engineering (SRE)**, proactive monitoring has become a crucial part of ensuring system reliability and meeting **Service Level Objectives (SLO)**. As we delve deeper into this chapter, we will explore how you can implement proactive monitoring in your Kubernetes clusters and service mesh to ensure reliable and high-quality services.

Importance and benefits of proactive monitoring

In the DevOps culture where teams take full ownership of the services they build, from design and development to deployment and maintenance, proactive monitoring is essential. This operational philosophy of *you build it, you maintain it* necessitates a deep understanding of your system's health and performance. It is here where proactive monitoring comes into play, enabling teams to stay ahead of issues and maintain the high availability and reliability of their services. Proactive monitoring offers several key benefits such as the following:

- **Early detection of issues**: Through continuous monitoring of your system's key metrics and logs, you can identify subtle changes that might indicate a potential problem. Early detection gives you the chance to resolve issues before they escalate into more significant problems or outages. For example, if you can detect a hard drive growing in size, you will be able to proactively remove some files or increase the size before the disk fills up.
- **Improved system reliability**: By identifying and resolving issues early, you can prevent them from causing service degradation or outages, improving the overall reliability of your system. Using the same example as above, the system was not impaired due to a full hard drive so our system remains reliable with no impact.
- **Informed decision-making**: Proactive monitoring provides a wealth of data that can help inform decision-making. You can use this information to optimize resource usage, plan capacity, prioritize development efforts, and more. Again, using our hard drive example, knowing the existing hard drive capacity will help the architects make appropriate changes in the capacities of the hard drives their purchase for new systems.
- **Enhanced customer satisfaction**: A reliable system that is free from frequent issues or downtime leads to better user experiences, which can significantly improve customer satisfaction. Our more reliable systems provide a better user experience and more satisfied customers!
- **Reduced operational costs**: While setting up a proactive monitoring system can require some upfront investment; it can significantly reduce costs in the long run. By preventing major issues and outages, you can avoid the substantial costs associated with downtime, including lost revenue and damage to your brand reputation.
- **Fostering a culture of accountability**: With proactive monitoring, first responders are equipped to be the first to know there is a problem by whatever alerting system is in place. For example, a paging system like PagerDuty which will alert the on-

call team directly. This sense of ownership fosters a culture of accountability within the team.

Proactive monitoring is not just an optional extra in today's complex multi-cloud, service mesh-enabled Kubernetes environments; it is a necessity. In the following sections, we will delve deeper into how you can implement proactive monitoring in your environments and the tools and practices that can help you achieve it.

Challenges in proactive monitoring

Implementing proactive monitoring in a complex system, such as a multi-cloud, service mesh-enabled Kubernetes environment, presents several challenges. Here, we will discuss some of these challenges and provide strategies on how to overcome them:

- **Data overload**: In comprehensive environments, proactive monitoring can generate an enormous amount of data, which can be overwhelming and can make it difficult to identify real issues amongst the noise. Imagine a fleet of 100 web servers with 10 different metrics all paging at the same time. This is also referred to as **Pager Fatigue**. This challenge can be addressed by setting the right thresholds for alerts and using anomaly detection algorithms that can highlight unusual patterns.

- **Silos and lack of visibility**: In a multi-cloud environment, monitoring data may become siloed within different clouds or different services. This lack of visibility can make it hard to detect problems that affect multiple components. Overcoming this challenge requires an integrated monitoring approach that brings together data from all sources into a single, centralized dashboard with your highest priority metrics.

- **Dynamic nature of Kubernetes**: The ephemeral nature of Kubernetes, with its dynamically scheduled and auto-scaling pods, can make traditional monitoring approaches ineffective. Tools specifically designed for Kubernetes can help, as they understand its architecture and can track metrics at the pod, node, and cluster level.

- **Complexity of service mesh**: Service mesh adds another layer of complexity to monitoring, as it involves monitoring not just the services themselves but also the mesh's control and data planes. Istio, for example, provides its own set of metrics that need to be monitored (refer back to the chapter covering service mesh for metrics). Tools that can integrate with service mesh and understand these metrics are necessary to address this challenge.

- **Tuning alert thresholds**: Too many alerts can lead to alert fatigue, while too few can result in missed problems. It is a challenge to tune the thresholds to get this balance right. Regular review and adjustment of alert thresholds, based on experience and changing conditions, can help maintain the right balance.

By recognizing and addressing these challenges, you can implement effective proactive monitoring that supports the DevOps philosophy of *you build it, you maintain it*, empowering your first responders to be the first to detect and respond to any issues.

Tools and techniques for proactive monitoring in Kubernetes

Monitoring your Kubernetes cluster proactively is critical for maintaining a healthy system and for following the DevOps mantra of *You build it, you maintain it*. The tools and techniques outlined below are intended to enable your first responders to detect, diagnose, and resolve issues before they escalate. This is by no means a complete list as there are many options. As always, list out your requirements and select a solution which solves them and your business use cases:

- **Prometheus**: This open-source tool has become the de-facto standard for monitoring Kubernetes environments. It provides detailed insights into your cluster's performance by scraping metrics from your services and storing them for analysis. It is also highly scalable and has a multi-dimensional data model with powerful queries. Check out **https://prometheus.io/** for more information and applicable use cases.
- **Grafana**: Often used in conjunction with Prometheus, Grafana is a data visualization tool that turns collected metrics into insightful graphs and dashboards, making it easier to spot trends or issues. The Grafana web page (**https://grafana.com/**) has a huge number of resources so check it out.
- **Jaeger and Zipkin**: These tools are used for tracing, in order to help diagnose and troubleshoot latency problems in microservices architectures. They provide visibility into the request flow across multiple services. Refer to **https://www.jaegertracing.io/** and **https://zipkin.io/** for how these tools interact with others in this list and common use cases.
- **Fluentd or Fluent Bit**: These are logging tools that can collect logs from different parts of your cluster and aggregate them into a single location. This makes it easier to search and analyze log data. Their respective web sites (**https://www.fluentd.org/** and **https://fluentbit.io/**) will have more information on how they fit into a monitoring stack.
- **AlertManager**: This tool works alongside Prometheus. It handles alerts sent by Prometheus, deduplicates, groups, and routes them to the correct receiver such as email, PagerDuty, or OpsGenie. Read more here: **https://prometheus.io/docs/alerting/latest/alertmanager/**.
- **Istio Metrics**: When using a service mesh like Istio, it provides its own set of metrics, giving insights into the behavior of the mesh's control and data planes. Please refer back to the previous chapter which covers Istio metrics.

- **Elastic Stack: Elasticsearch (ELK)** for search, Logstash for centralized logging and Kibana for visualization. This stack helps in logging infrastructure.

In terms of techniques, let us consider a few:

- **Implementing proper log management**: Create consistent log messages across services to make it easier to search and analyze logs. This allows for all stakeholders and consumers of the logs to understand them better and have trust in the reliable system.
- **Creating effective dashboards**: Design dashboards that quickly provide the status of your system. They should be easily understandable, and it should be straightforward to drill down to get more detail. Dashboards are a great way to be transparent and build trust with your stakeholders.
- **Setting intelligent alerts**: Too many alerts can cause alert fatigue, while too few might miss important events. Set alerts on conditions that indicate systemic problems rather than individual service failures. Actionable alerts build trust and do not waste anyone's time.
- **Automating incident response**: Use tools that can automatically mitigate the known issues. Automated responses can often prevent minor issues from escalating. When there is a repeatable and consistent process it is easier to follow and is better trusted by the team.

By using these tools and techniques, you can ensure that your first responders are well-equipped to maintain system health and performance.

Implementing proactive monitoring

Implementing proactive monitoring involves multiple steps, each of which contributes to maintaining a healthy system and adhering to the DevOps principle of *you build it, you maintain it*. We are going to deep dive into this topic so the reader may have a solid understanding of this critical process.

Understanding your system

A deep understanding of your system is a fundamental prerequisite to implementing effective proactive monitoring in any environment, whether it is a single cluster, multi-region, or multi-cloud Kubernetes environment. The entire system could comprise of the underlying infrastructure or platform, such as Kubernetes. The application architecture along with the code logic and behaviors are important to understand as well as the interfaces to other systems. Your complete understanding of the system and the environment in which it operations will serve your well.

Problem it solves

Understanding your system helps you identify what is 'normal' for your application and infrastructure. Without a clear understanding of your system's usual behavior, you cannot accurately define what constitutes an anomaly or a system failure. You need to understand what metrics to collect, what values are normal for those metrics, and at what point you should be alerted because the system is not behaving as expected.

Additionally, understanding your system helps you identify dependencies between services or applications, which is crucial for root cause analysis. If a service fails, it is important to know what other services might be affected as a result.

Solving the problem

There are three setups:

- **Single Kubernetes cluster**: Start by understanding the architecture of your application. What services does it contain? How do they interact? Next, understand the performance characteristics of your services. What is their typical CPU and memory usage? What is the average, minimum, maximum, and standard deviation of their response time? Use monitoring tools to collect these metrics and observe them over time to establish a baseline.
- **Multi-region Kubernetes clusters**: In addition to understanding individual clusters, you need to understand the interactions between clusters in different regions. How do they communicate? What is the latency between them? How does failover work? Moreover, do consider region-specific issues, like the impact of regional outages and latency differences.
- **Multi-cloud Kubernetes clusters**: Multi-cloud environments add another layer of complexity because you are dealing with different cloud providers' services, each with their own quirks and performance characteristics. Understand how your clusters are distributed across cloud providers, how they interact, and how your application behaves when there are inter-cloud communication issues.

By comprehensively understanding your system in these environments, you can set up a proactive monitoring strategy that captures the right metrics, sets meaningful alerts, and thus helps maintain the reliability of your applications and infrastructure. It enables you to get ahead of issues before they become serious problems and allows you to deliver on the DevOps promise, *you build it, you maintain it*.

Selecting the right tools

Proper tool selection is a key aspect of setting up proactive monitoring across any Kubernetes environment, be it a single cluster, multi-region, or multi-cloud setup. Refer to the following figure:

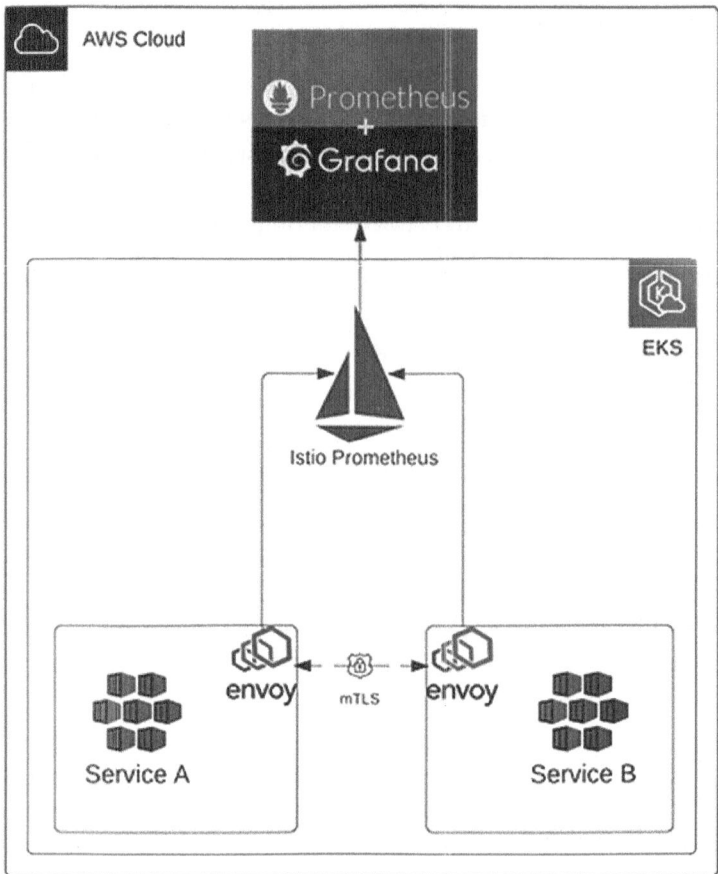

Figure 10.1: Istio service mesh with Prometheus and Grafana

Problem it solves

The right monitoring tools will allow you to accurately and efficiently gather and analyze data from your applications and infrastructure, providing the insights you need to identify and rectify issues before they negatively impact your system. Without suitable tools, you risk collecting insufficient or irrelevant data, leading to longer downtimes, slower resolution times, and overall poorer system performance and user experience.

Solving the problem

Let us go over the different setups:
- **Single Kubernetes cluster:** A single cluster might be well-served by a simpler monitoring tool stack, such as Prometheus for metric collection, Grafana for visualization, and Alert manager for alerts. These tools are open-source and widely used, and they integrate well with Kubernetes.

- **Multi-region Kubernetes clusters:** When you have multiple clusters across different regions, it is important to have centralized monitoring that aggregates metrics from all regions. This might involve using more sophisticated commercial tools like Datadog or New Relic, or setting up a global view in Grafana that pulls data from Prometheus servers in each region.
- **Multi-cloud Kubernetes clusters:** In a multi-cloud environment, you will need to deal with the different monitoring services provided by each cloud provider. A tool like Prometheus could still be used for in-cluster metrics, but for cloud-specific metrics, you might need to pull data from Amazon CloudWatch, Google Cloud Monitoring, Azure Monitor, and so on. An open-source tool like OpenTelemetry can help gather tracing and metrics data across all your services, regardless of where they are hosted.

Choosing the right tools for your specific environment and needs will enable you to effectively implement proactive monitoring. This ensures you have the correct data to make informed decisions about your system's health and to identify potential problems before they affect your users, allowing your team to honor the DevOps adage of *you build it, you maintain it*. Remember that the goal is not to collect the most data but to collect the right data.

Configuring the tools

Once you have selected the appropriate monitoring tools, the next step is to configure them correctly, which is critical regardless of your Kubernetes setup, whether it is single cluster, multi-region, or multi-cloud.

Problem it solves

Proper tool configuration ensures that you are monitoring the right aspects of your infrastructure and applications and that the data collected is accurate, relevant, and insightful. If your tools are not configured correctly, they may provide misleading information, miss crucial data, or generate unnecessary noise, resulting in an ineffective monitoring system and potential unnoticed issues. For example, when configuring a hard drive limit setting and setting the threshold too high, your team may only have a short time to remediate versus a more appropriate value which would have allowed them more time to react.

Solving the problem

Let us go over the different setups:
- **Single Kubernetes cluster**: For a single cluster, configuration might be simpler. It involves setting up the tools to monitor your workloads, the Kubernetes system, and any underlying infrastructure. This could include configuring Prometheus to

scrape metrics from your services or setting up alerting rules in Alert manager to notify you of any abnormal conditions.

- **Multi-region Kubernetes clusters**: For multi-region clusters, you need to ensure that each cluster's metrics are collected and can be differentiated. You might need to add extra labels in Prometheus for cluster identification, set up federation to aggregate metrics from different regions, or configure your tool to send metrics to a central location for a global view.

- **Multi-cloud Kubernetes clusters**: Configuration complexity increases in a multi-cloud environment. You have to accommodate the different services, standards, and tools of each cloud provider. For example, you might need to configure the AWS CloudWatch agent for EKS clusters, the Stackdriver agent for GKE clusters, and so on. You also need to ensure that all these different metrics can be aggregated into a meaningful, coherent view.

Overall, configuring your tools correctly ensures that your monitoring system can accurately represent the state of your system. This allows you to quickly and reliably detect issues and anomalies, ensuring that the first team to know about any problems is your first responders, as per the DevOps way of *you build it, you maintain it*. This also helps to provide a proactive response, minimizing downtime and maximizing system performance.

Building dashboards

Dashboarding is a key step in implementing proactive monitoring. Dashboards provide a visual representation of the data collected by your monitoring tools, which allows teams to understand the state of their systems at a glance.

Problem it solves

Monitoring a complex system like Kubernetes generates a lot of data. Without an effective way to understand and interpret this data, it can be overwhelming and obscure potential issues. Dashboards solve this problem by visualizing the data in an intuitive and actionable way. For example, having dashboards for individual systems or customers allows you to see patterns you may have missed if viewing all metrics combined. They make it easier to identify patterns, understand trends, spot anomalies, and track the impact of changes or incidents.

Solving the problem

Let us go over the different setups:

- **Single Kubernetes cluster**: In a single cluster, a dashboard can provide a comprehensive view of the cluster's state. This includes metrics like CPU usage, memory usage, network traffic, and pod status, among others. Tools like Grafana

are commonly used to build these dashboards, with data sourced from monitoring tools like Prometheus.
- **Multi-region Kubernetes clusters**: In a multi-region setup, it is beneficial to have both global dashboards that show the health and performance of your entire system, and individual dashboards for each cluster. This allows you to monitor the system at a high level, while also being able to dive into specifics when needed. It is also beneficial to incorporate geolocation data, so you can easily see which regions are experiencing issues.
- **Multi-cloud Kubernetes clusters**: In a multi-cloud environment, building dashboards becomes more complex but also more critical. It is essential to have a unified dashboard that can aggregate and display data from all clusters, regardless of the cloud provider. This may involve normalizing metrics across different cloud providers and making sure the dashboards handle the different types of metrics each provider offers.

By building effective dashboards, you enable your first responders to quickly identify and investigate issues. This aligns with the DevOps principle of *You build it, you maintain it*, by giving the team responsible for building the system the tools they need to maintain it effectively. Dashboards provide a visual representation that can make it easier to detect issues before they affect stakeholders, contributing to the goal of proactive monitoring.

Setting up alerts

Setting up alerts is a crucial component of proactive monitoring. Alerts notify first responders of potential issues within your system based on predefined criteria or anomalies detected in the system's performance metrics. For example, a slow web service affecting customer satisfaction or a database with full disks. These need to be addressed quickly and require the alert of the on-call team.

Problem it solves

Monitoring systems continuously generate vast amounts of data. Sifting through all this data manually to identify potential issues is not feasible, especially in real-time. Without alerts, teams may not become aware of a problem until it has already affected the system's performance or availability, which could lead to downtime and service disruption.

Solving the problem

Let us go over the different setups:
- **Single Kubernetes cluster**: In a single-cluster environment, alerts can be set up to monitor various aspects of cluster health, including CPU and memory usage, network traffic, and pod status. For instance, an alert can be triggered when CPU usage exceeds a certain threshold, indicating potential resource exhaustion. Alerts

are typically set up using tools like Prometheus and sent to an incident response platform or via email or slack.

- **Multi-region Kubernetes clusters**: With multi-region clusters, alerts become even more important, as issues may arise in one region that do not affect others. Alerts can be configured based on region-specific metrics and should be routed to the appropriate teams responsible for each region.
- **Multi-cloud Kubernetes clusters**: In a multi-cloud environment, alerts should be set up for each cloud provider's specific metrics, as well as for metrics that are common across providers. Alerts might also need to take into account inter-cloud communication issues, such as latency or data transfer costs.

In all cases, alerts should be meaningful and actionable. Too many alerts, especially false positives, can lead to alert fatigue and desensitize teams to potential issues. Alert thresholds and conditions should be continually reviewed and adjusted as necessary to maintain their relevance.

Alerting is essential in the *you build it, you maintain it* DevOps mindset because it ensures that the teams responsible for building the system are also the first to know when issues arise. By setting up effective alerts, first responders can identify and address problems before they affect stakeholders, thus achieving the goal of proactive monitoring.

Testing your setup

Testing your monitoring setup is a critical part of the implementation process. The main objective here is to verify that your monitoring system can indeed detect problems and that alerts are triggered as expected. Hosting fire drills, where you intentionally trigger and alert, can help the team practice repeatedly and hone their troubleshooting skills. This also allows teams to perfect their SOP's to be more efficient. This helps ensure that when a real issue occurs, your system will notify the right people with accurate information about the problem.

Problem it solves

Without testing, you run the risk of having a monitoring system that fails to notify you when something goes wrong. It is like having a fire alarm that does not sound when a fire breaks out. Furthermore, untested alert thresholds could result in too many or too few alerts, both of which can lead to problems being missed or ignored.

Solving the problem

Let us go over the different setups:

- **Single Kubernetes cluster**: Testing in a single cluster environment involves triggering situations that should raise alerts. For example, you can artificially increase the CPU or memory usage in a pod to ensure the alert is triggered. This

process helps ensure that the alert system works as expected and the notifications reach the intended recipients.
- **Multi-region Kubernetes clusters**: In a multi-region environment, testing becomes a bit more complex as it must ensure that region-specific issues are accurately detected and that alerts are routed correctly. This might involve simulating a regional outage or network latency.
- **Multi-cloud Kubernetes clusters**: Testing in a multi-cloud environment involves validating that the monitoring system can accurately track metrics across different cloud providers and that cloud-specific issues trigger the appropriate alerts. This might mean simulating cloud-specific scenarios like a provider outage or connectivity problems between clouds.

In all these cases, the testing phase is not a one-off process but should be repeated whenever changes are made to the system or the monitoring setup.

In the DevOps way of *you build it, you maintain it*, testing your setup ensures that your first responders have the tools they need to detect and respond to issues promptly. By verifying that your monitoring and alerting system works correctly, you can avoid unpleasant surprises and make sure problems are detected before your stakeholders become aware of them.

Iterative improvement

Iterative improvement is a continuous process that involves regularly reviewing, updating, and refining your monitoring setup. It encompasses understanding what metrics useful, fine-tuning alert thresholds are, and adjusting dashboards to reflect the evolving needs and realities of your system. All too often monitoring configuration, alerts and other components are forgotten and stop producing meaningful alerts. Without regular maintenance and upkeep, a monitoring system can become ineffective.

Problem it solves

Without iterative improvement, your monitoring setup can become outdated, leading to a decreased visibility into your system's performance and health. As your application evolves, so will the potential issues that can arise. Without updating your monitoring setup, you risk missing crucial information or receiving false alerts that could undermine the effectiveness of your monitoring system.

Solving the problem

Let us go over the different setups:
- **Single Kubernetes cluster**: In a single cluster environment, iterative improvement might involve adjusting metrics and alerts as your application evolves, scaling up, or adding new features. For example, as the load on your application grows,

you may need to revise alert thresholds or add new metrics to ensure you're still effectively monitoring the health of your system.

- **Multi-region Kubernetes clusters**: For multi-region setups, iterative improvement includes refining your setup to accommodate for the differences between regions, such as latency or regional traffic patterns. Over time, as you understand the unique traits of each region better, you can tailor your monitoring and alerting setup to better fit these nuances.
- **Multi-cloud Kubernetes clusters**: Iterative improvement in a multi-cloud setup involves updating your monitoring configuration to handle cloud-specific metrics and alerts. Each cloud provider may offer unique metrics or have specific characteristics that you need to account for in your monitoring setup. As you gain more experience with each provider, you can continuously refine your monitoring setup to ensure optimal visibility.

Iterative improvement embodies the DevOps principle of *you build it, you maintain it*, emphasizing the need for ongoing maintenance and refinement of your systems. By regularly reviewing and updating your monitoring setup, you ensure that your first responders always have the most accurate and useful information to prevent and respond to issues, thereby maintaining the trust of your stakeholders.

By following these steps, you can ensure that your first responders have the information they need to detect and address issues before they impact your stakeholders. Proactive monitoring takes effort to set up and maintain, but the payoff in terms of system availability and performance is well worth it.

Understanding and using metrics, logs, traces and alerts

As part of the DevOps practice where the builder of the system also maintains it, comprehensive monitoring is vital. Early detection of problems not only prevents cascading issues that could bring the system down but also saves time and resources that would be spent on remediation. Let us delve into the key components of monitoring: metrics, traces, logs, and alerts.

Refer to the following figure:

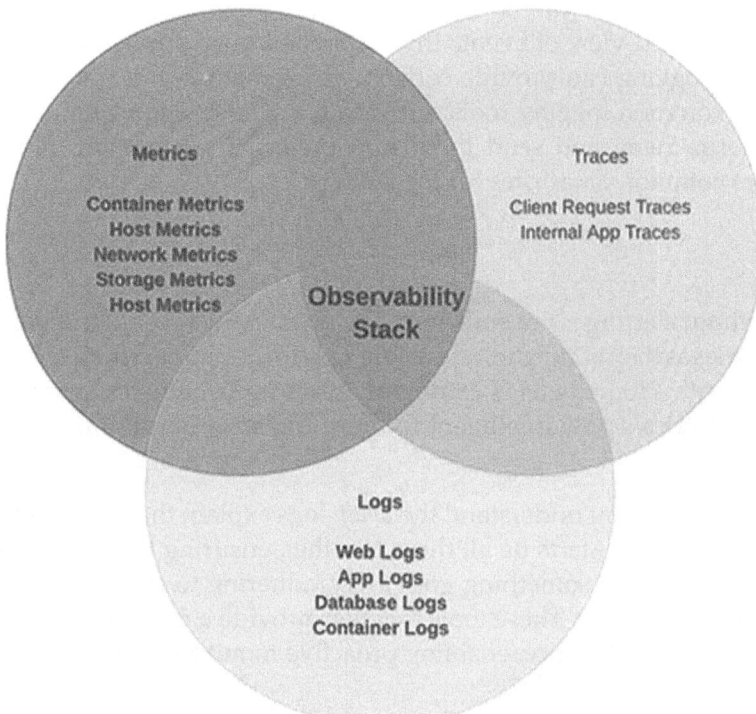

Figure 10.2: Metrics, Traces and logs

Metrics

Metrics provide a numeric representation of data measured over intervals of time. They play a crucial role in understanding the behavior of our system, help identify unusual patterns, and are fundamental to observing a Kubernetes environment. Key Kubernetes metrics include resource utilization (like CPU and memory usage), network usage, and disk I/O, among others. Tools such as Prometheus can be used to collect these metrics, and Grafana can visualize them.

Traces

Tracing captures the lifecycle of a request as it travels through the various services of an application. Distributed tracing, in particular, is vital for debugging and monitoring distributed systems like Kubernetes and Istio. It enables developers to track a request as it travels through the microservices architecture, helping them understand latency issues and identify bottlenecks. Tools like Jaeger and Zipkin provide distributed tracing capabilities.

Logs

Logs provide a detailed view of events that happened in an application or a service. In a complex system, logging can provide context where metrics and traces cannot. Fluentd or Logstash are common logging tools, providing a way to aggregate logs from various sources, transform them, and send them to a preferred destination. Elasticsearch and OpenSearch are common datastores for log storage.

Alerts

Monitoring without alerting is not sufficient. The primary purpose of monitoring is to raise awareness of issues as they occur and, if possible, before they impact the users. AlertManager in combination with Prometheus is commonly used in Kubernetes environments. It can digest complex metrics, apply intelligent routing, and send notifications through various channels.

In essence, metrics help you understand the *what*, logs explain the *why*, and traces uncover the *where* in your system. Alerts tie all these together, ensuring that your first responders are the first to know when something goes awry, adhering to the core DevOps principle of *you build it, you maintain it*. These tools together provide a comprehensive view of your system's health and performance, enabling proactive monitoring and timely resolution.

Best practices for proactive monitoring

After covering the benefits of a proactive monitoring system, we begin to plan for implementing one. In order to effectively implement proactive monitoring in your Kubernetes environment, the following best practices should be considered:

- **Establish key performance indicators (KPIs)**: KPIs will provide clear expectations of what success looks like in your system. They are quantifiable goals that are vital to the health of your system. Examples can include system availability, request response times, error rates, or resource usage rates.
- **Comprehensive monitoring**: Cover all aspects of your system with monitoring. This includes the applications, infrastructure, network, and services, among others. Metrics, traces, and logs should be collected from all these sources for a comprehensive view of the system health.
- **Set meaningful alerts**: While alerting is crucial, too many alerts can lead to alarm fatigue and might lead to important alerts being overlooked. Prioritize and classify alerts based on their importance and impact on the system.
- **Use auto-scaling**: Kubernetes has built-in auto-scaling capabilities which can be leveraged based on the metrics collected. This not only helps in maintaining system performance but also can be cost-efficient.

- **Monitor your monitoring system**: Even the best monitoring systems can fail. It is important to monitor your monitoring system itself and have a backup plan to prevent a single point of failure.
- **Regular review**: Continually review your monitoring setup. As your system evolves, your monitoring needs might change too. It is important to align your monitoring setup with your current system needs.
- **Training**: Your team should be well trained not only to interpret the data collected but also to react effectively. A well-monitored system can only be as effective as the team that operates it.
- **Leverage AI for IT operations (AIOps)**: AIOps uses machine learning and data science to simplify and enhance IT operations. It helps in noise reduction, anomaly detection, root cause analysis, and predictive analytics, enhancing the proactive monitoring capabilities.

Remember, the key to successful proactive monitoring is not just to collect data but to use that data to gain actionable insights. The goal is to identify potential issues before they impact the system, maintain system health, and improve overall system performance.

Securing your monitoring stack

Securing your monitoring stack is just as critical as setting it up. An exposed or vulnerable monitoring system could provide a backdoor into your environment, undermining the security and stability of your entire operation. Here are some key considerations and best practices to ensure the security of your monitoring stack:

- **Data encryption**: Both data-in-transit and data-at-rest should be encrypted. Using encryption for data-in-transit ensures that sensitive data cannot be intercepted during transfer. For data-at-rest, encryption protects the stored data from unauthorized access.
- **Access control**: Ensure that only authorized individuals have access to the monitoring data and tools. Use **Role-Based Access Control (RBAC)** to assign permissions and limit access based on individual roles and responsibilities.
- **Audit logs**: Keep track of who does what in your monitoring system. Audit logs help you track all activity, which can be critical for identifying misuse, understanding incidents, and maintaining regulatory compliance.
- **Secure communication**: Ensure that all communication between your services is secure. Consider using **mutual Transport Layer Security (mTLS)** for service-to-service communications to provide two-way verification of security credentials.
- **Patch management**: Regularly update and patch your monitoring tools to protect against known vulnerabilities.

- **Network policies**: Use Kubernetes network policies to control the traffic between your pods and between your services. This can prevent unauthorized access to your monitoring stack.
- **Secret management**: Use secure methods to manage your secrets, such as API keys, tokens, and certificates. Kubernetes provides a Secret resource, but other tools like HashiCorp's Vault can provide more robust solutions. You can also rely on and integrate cloud based key stores however you will run into issues with multi-cloud Kubernetes deployments. It is recommended to have a cloud agnostic solution here.
- **Incident response**: Have a plan in place for when a security incident occurs. This should include identifying the issue, containing the damage, eradicating the cause, recovering, and conducting a post-mortem to learn from the incident.

Remember, security is not a one-time setup but an ongoing process. Regular reviews and updates to your security practices, as well as continuous monitoring for potential security threats, are vital to the integrity of your monitoring stack.

Proactive monitoring in service mesh

As organizations increasingly adopt microservices architecture, the complexity of managing communication and enforcing policies among these services can be daunting. Service meshes, like Istio and Envoy, offer a solution for handling this complexity in a more organized and efficient manner. However, in order to optimize their functionality and ensure smooth running of the services, proactive monitoring becomes crucial.

In the context of a service mesh, proactive monitoring involves the continuous observation and analysis of the status, performance, and interactions of the mesh components and the microservices running within it. This includes collecting metrics, logs, and traces from both the service mesh control plane and data plane to gain visibility into the health and performance of your services.

Metrics help you understand the quantitative state of your service mesh. They include data on request rate, error rate, duration of requests, etc. Istio, for instance, provides a rich set of metrics via Prometheus, which can be visualized using tools like Grafana.

Logs provide qualitative information on what is happening within your service mesh. This can include data about specific events, errors, or other informative messages. Envoy provides extensive logging capabilities, which can be collected and analyzed with a centralized logging system like ELK stack.

Traces provide an insight into how requests propagate through your services and are especially useful in identifying bottlenecks and latency issues. Istio integrates with distributed tracing systems like Jaeger and Zipkin, which enable you to track a request as it travels through various microservices in the mesh.

Setting up alerting is another crucial aspect of proactive monitoring. This involves configuring rules to trigger notifications when certain conditions are met, such as an unexpected increase in error rates or latency. This helps ensure you can respond quickly to potential issues and prevent them from escalating.

As with any monitoring setup, security is paramount. This includes securing the communication between your monitoring tools and your service mesh, ensuring only authorized users have access to your monitoring data, and keeping your monitoring tools updated to prevent potential vulnerabilities.

Proactive monitoring in the context of a service mesh is about maintaining visibility into your microservices environment. By continuously collecting, analyzing, and acting on monitoring data, you can ensure the smooth operation of your services, swiftly address issues as they arise, and maintain a high quality of service for your users.

Conclusion

We have embarked on an important journey through the world of proactive monitoring in Kubernetes, primarily focusing on the importance and benefits, challenges and their possible solutions, tools and techniques, implementation strategies, metrics and alerts usage, best practices, and security considerations.

Throughout this chapter, we continuously emphasized the DevOps principle, *you build it, you maintain it*, and reiterated the importance of first responders being the first to know that there is an issue. We have also delved into how to implement and use metrics, logs, traces, and alerts effectively for proactive monitoring to ensure smooth and reliable operation of Kubernetes clusters.

Furthermore, we took a deep dive into the application of proactive monitoring within the context of a service mesh. Here, we explained how the principles of proactive monitoring apply when you are using a service mesh like Istio and Envoy, and how metrics, logs, traces, and alerting can be effectively used within this context to maintain high service quality and rapidly address issues.

In the next chapter, get ready to take your skills to the next level as we will learn about enabling comprehensive observability. This exciting next step will teach you how to further improve the reliability and efficiency of your Kubernetes clusters and applications. Get ready to discover new ways to increase visibility into your systems, enhance diagnostics, and more. Keep going, you are doing great!

Join our book's Discord space

Join the book's Discord Workspace for Latest updates, Offers, Tech happenings around the world, New Release and Sessions with the Authors:

https://discord.bpbonline.com

CHAPTER 11
Enabling Comprehensive Observability

Introduction

In this era of increasingly complex and distributed systems, monitoring the health and performance of your applications and infrastructure is not sufficient. The advent of microservices architecture and multi-cloud Kubernetes deployments necessitate a broader view, one that encompasses not just what is happening inside your system but also how it is behaving in the context of its external environment. This is where comprehensive observability comes in.

In this chapter, we will delve into the concept of comprehensive observability, which goes beyond conventional monitoring to provide a holistic and detailed perspective of your infrastructure and applications' health on a global scale.

Structure

Here is an overview of the topics we will cover:
- Introduction to comprehensive observability
- Importance of observability in multi-cloud Kubernetes
- Understanding the three pillars of observability
- Tools and techniques for achieving comprehensive observability

- Implementing observability in a multi-cloud Kubernetes environment
- Observability with service mesh
- Best practices for comprehensive observability
- Challenges in implementing observability and how to overcome them
- Case studies: Observability in action

Objectives

By the end of this chapter, you will gain a thorough understanding of comprehensive observability and why it is critical for managing multi-cloud Kubernetes deployments. You will get acquainted with the three pillars of observability, that is, logs, metrics, and traces, and learn how to leverage them for in-depth insight into your systems. We will explore the tools and techniques that will enable you to achieve comprehensive observability, focusing on their practical implementation in a multi-cloud Kubernetes environment. This chapter will also cover the importance of observability in the context of a service mesh, share best practices, and address the common challenges in implementing observability, equipping you with the knowledge to navigate and overcome them. By examining real-world case studies, you will see how observability is applied in various scenarios, providing valuable insights for your unique context. Get ready to take your operational insight to the next level with comprehensive observability.

Introduction to comprehensive observability

As businesses scale, they often adopt microservices architecture and expand their operations across multiple cloud providers. With this distributed system, the complexity to maintain and monitor the entire operation also increases. Hence, it is critical to have a comprehensive observability mechanism to understand the health and performance of your systems, clusters, infrastructure, and applications at a global level.

Comprehensive observability is not just about monitoring individual system metrics. Instead, it provides a holistic view of your system's health and performance by aggregating and analyzing data from multiple sources, including logs, metrics, and traces, often referred to as the three pillars of observability. These data points provide detailed insights into the state and behavior of your systems and applications, whether they are running on a single cluster, multi-region, or multi-cloud Kubernetes environment.

With comprehensive observability, you can detect, diagnose, and resolve issues faster and more effectively. It provides context-rich information that helps you understand how individual components affect overall system performance. It helps in identifying patterns, trends, and anomalies, making it easier to predict and prevent potential issues before they impact users.

For instance, in a multi-cloud Kubernetes environment, each cloud provider's infrastructure might behave slightly differently due to the differences in their underlying architecture. For example, AWS utilizes security groups vs. Azure AKS and Google GKE use network policies, this can affect how you define and manage network access for your Pods. Another important example is the various load balancers options and their integration with their respective network solutions. Control plane management, node pools, scalability algorithms and many more resources behave and are configured differently in each cloud. By implementing comprehensive observability, you can monitor these nuances and adjust your operations accordingly to maintain optimal performance across all environments.

In essence, comprehensive observability is not just a tool; it is a philosophy. It emphasizes proactive monitoring, root cause analysis, and continuous improvement to drive system reliability and resilience. In the upcoming sections, we will delve deeper into the benefits, challenges, tools, and best practices for implementing comprehensive observability in multi-cloud Kubernetes environments.

Importance of observability in multi-cloud Kubernetes

As enterprises continue to shift towards multi-cloud Kubernetes environments, the demand for advanced, robust, and comprehensive observability grows. Observability, in this context, extends beyond monitoring and involves gaining deep insights into the system's behavior and state. The goal is not only to understand what is happening but also why it's happening.

Let us delve into why observability is crucial in multi-cloud Kubernetes environments.

Complexity and variability

Multi-cloud Kubernetes environments are complex and highly variable by nature. You may have different services running on different cloud providers, each with its own peculiarities and management systems. An example is an e-commerce platform split across AWS EDS for the front-end website, Google GKE for the product catalog, and Azure AKS for payment processing. The intricate interactions between these services can often lead to unforeseen issues. Comprehensive observability allows you to make sense of these complexities and variabilities, providing a unified view of your entire system's performance across multiple clouds.

Troubleshooting and root cause analysis

When problems arise in a multi-cloud environment, pinpointing the root cause can be like finding a needle in a haystack. With comprehensive observability, you can quickly identify where things are going wrong, reducing the time spent on troubleshooting and allowing for quicker problem resolution.

Performance optimization

Observability helps you understand the performance characteristics of your applications and infrastructure across different cloud providers. Using the previous split e-commerce example; you would measure the page load times, request latency, error rates and resource utilization on AWS for the front-end website while you would observe the database query performance, cache hit rates, API response times and availability of the product catalog running in GKE. You can use this information to optimize resource utilization, identify performance bottlenecks, and make data-driven decisions to improve overall system performance.

Proactive problem detection

Comprehensive observability can help you detect anomalies and potential issues before they escalate and affect the user experience. Continuing on with our split ecommerce example from before; you would be on the lookout for sudden spikes in page load times or sudden increases in HTTP 500 errors from the front-end website on AWS while looking for frequent database timeouts and high cache miss rates on the product catalog running on GKE. This proactive approach can prevent downtime and contribute to maintaining high levels of system availability and reliability.

Insight and understanding

Observability provides valuable insights into the behavior of your systems. Understanding how your applications perform under various conditions can inform future development and operational decisions, leading to more resilient and efficient systems.

As an example, imagine you run a small online music store:
- Your website is hosted on AWS EKS, and you use observability tools to monitor page load times across different web pages.
- You notice a specific product page consistently loads slower than others.
- Digging deeper, you discover:
  - This page displays several large music sample files for each product.
  - Downloading these samples is causing the slow page load.

- Based on this insight, you can take the following actions:
    - Offer smaller sample files or implement lazy loading, where samples only download when clicked.
    - This improves the overall website performance and user experience.

This is a simplified example, but it demonstrates how observability provides concrete insights (slower page load linked to specific samples) that fuel better decisions (optimize sample handling) to build a more efficient system (faster website).

Cost management

Multi-cloud environments can quickly become cost-intensive, especially if resources are not optimized. Observability can provide insights into resource usage patterns, enabling more effective cost management and optimization. Imagine you run a blog hosted across two cloud providers:

- **Front-end website hosted on Google Kubernetes Engine (GKE):** It handles user traffic and displays blog posts.
- **Database hosted on Microsoft Azure:** It stores all blog content and comments.
- You use observability tools to track your database resource usage:
    - You discover low utilization during off-peak hours (for example, late nights).
    - You also see spikes in usage during peak times (for example, evenings).
- Based on this insight, you can take the following actions:
    - Implement autoscaling on your Azure database cluster.
    - This automatically scales down resources during off-peak hours, reducing your compute costs.
    - During peak times, the cluster automatically scales up to handle increased traffic, maintaining performance.

Outcome: Observability helps you understand resource usage patterns (low off-peak, high peak) and implement cost-effective solutions (autoscaling) to optimize your cloud spending without compromising performance.

This is a basic example, but it demonstrates how observability can guide data-driven cost management decisions in multi-cloud environments. By observing your resource usage, you can identify opportunities for optimization and avoid unnecessary spending.

In summary, comprehensive observability in multi-cloud Kubernetes environments is not a luxury but a necessity. It is critical for maintaining system performance, reliability, and cost-efficiency in today's complex and highly variable cloud landscapes. In the next sections, we will explore the tools and practices that can help you achieve comprehensive observability in your multi-cloud Kubernetes environments.

Understanding the three pillars of observability

In the domain of system observability, three elements form the cornerstone, commonly known as the three pillars of observability. These are Logs, Metrics, and Traces. These pillars offer different perspectives and provide a holistic view of your systems. Let us discuss these pillars and their relevance in multi-cloud Kubernetes environments.

Logs

A log is a time-stamped record of discrete events that happened over time. Logs can provide information about what is happening in your system, often at a very granular level. For instance, an application log might contain information about the requests it is processing, any errors or warnings it encounters, or diagnostic information helpful for debugging. In a multi-cloud Kubernetes environment, you might have logs from your Kubernetes systems (for example, kubelet logs), application logs, and logs from your underlying cloud provider services. Aggregating and analyzing these logs can provide deep insights into your system's behavior and help you troubleshoot problems.

Metrics

Metrics are numerical representations of data measured over intervals of time. They are essential for understanding the general performance of your system. Examples of metrics include the number of requests per second your application is serving, the CPU usage of your nodes, or the latency of your services. In a multi-cloud environment, you may be collecting metrics about your Kubernetes systems (like pod memory usage), application performance metrics (like transaction times), and metrics from your cloud provider services (like storage IOPS). Metrics can help you identify trends, compare the performance of different systems, and set alerts on anomalous behavior.

Traces

Tracing involves tracking a request as it travels through your system and services, often represented as a directed acyclic graph of the path and timing of the request. Traces are particularly important in microservices architectures, where a single user request might interact with many services. In a multi-cloud Kubernetes environment, you might trace a request as it travels from a service on one cloud provider, through your service mesh, to a service on another cloud provider. Tracing can help you understand the flow of requests in your system, identify bottlenecks or failures, and troubleshoot complex issues.

By combining the data from logs, metrics, and traces, you can gain a comprehensive view of your systems' health and performance in your multi-cloud Kubernetes environments. Each pillar offers a unique perspective and, when used together, provides a multidimensional view of your system's behavior. In the following sections, we will delve into how to use these pillars effectively in multi-cloud Kubernetes environments.

Tools and techniques for achieving comprehensive observability

When it comes to achieving comprehensive observability in multi-cloud Kubernetes environments, numerous tools and techniques can be employed. Each tool or technique addresses a specific pillar of observability (logs, metrics, traces) or bridges multiple pillars. Here are some key tools along with their highlights.

Prometheus

A widely adopted tool for collecting and querying metrics. Prometheus allows you to measure various system parameters, from hardware utilization to application performance, and supports alerting based on these metrics. It is cloud-agnostic, making it suitable for multi-cloud environments.

Here are some key highlights:

- **Time-series database**: Stores metrics in a highly efficient and scalable way, enabling real-time performance monitoring.
- **Pull-based model**: Collects metrics directly from targets, reducing overhead and simplifying deployment.
- **Cloud-native ecosystem**: Integrates seamlessly with Kubernetes and other cloud-native technologies.

Grafana

This visualization tool is often used alongside Prometheus to create dashboards that can display metrics data in real-time. Grafana supports querying, visualizing, and understanding the metrics. It provides a high-level, global view of the system's health and performance.

Here are some key highlights:

- **Visualization and dashboarding**: Creates rich, interactive visualizations of metrics and logs, enabling comprehensive data exploration.
- **Extensive plugin ecosystem**: Supports various data sources and visualization types, customizing dashboards to specific needs.
- **Collaboration features**: Enables team members to share and annotate dashboards, enhancing communication and decision-making.

Elastic Stack

Elasticsearch is a search engine that can store and search logs, Logstash is a log pipeline tool that accepts inputs from various sources, and Kibana is a visualization tool for data in Elasticsearch. The ELK stack allows you to aggregate logs from different clusters and regions, providing a unified view of your logs.

Some key highlights are:

- **Centralized logging**: Collects, stores, and analyzes logs from diverse sources, providing insights into system behavior and troubleshooting issues.
- **Full-text search**: Enables powerful search capabilities across log data, pinpointing relevant information quickly.
- **Scalability and resilience**: Handles large volumes of data and supports high availability for mission-critical operations.

Jaeger and Zipkin

These tools are used for tracing requests in distributed systems. They help you track a request as it travels through various services in your system and provide insights into performance bottlenecks or failures.

Some key highlights are as follows:

- **Distributed tracing**: Track requests as they flow through multiple services, identifying performance bottlenecks and troubleshooting complex interactions.
- **Open-source standards**: Support OpenTracing and OpenTelemetry standards for compatibility across tools and vendors.
- **Visualization**: Provide intuitive UIs to visualize trace data and understand service dependencies.

Fluentd/Fluent Bit

These are log shippers that can send logs to various destinations. Fluentd and Fluent Bit are cloud-agnostic, supporting numerous inputs and outputs, and can aggregate logs from different sources into a unified view.

Some key highlights are:

- **Log collection and forwarding**: Collect logs from various sources and forward them to centralized logging systems for analysis.
- **Flexible configuration**: Support different filter plugins and output destinations, customizing data pipelines.
- **Lightweight and efficient**: Optimized for resource-constrained environments like Kubernetes.

Service mesh

A service mesh provides observability into your system by providing automatic telemetry, logging, and tracing capabilities. Istio and Envoy, for example, can generate logs, metrics, and traces for all traffic within the mesh.

Some key highlights are:

- **Service mesh**: Provide a layer of infrastructure for managing service-to-service communication, enabling observability features like distributed tracing and metrics collection.
- **Traffic management**: Control traffic flow, implement routing rules, and apply security policies for microservices.
- **Observability built-in**: Generate metrics and traces for service interactions, providing insights into network behavior and application performance.

OpenTelemetry

This project provides a set of APIs, libraries, agents, and instrumentation that can generate and collect telemetry data (metrics, logs, and traces) from your services. OpenTelemetry is designed to be platform-agnostic, making it an excellent fit for multi-cloud environments.

Some key highlights are:

- **Open standard for telemetry**: Establishes a vendor-neutral API and SDKs for collecting metrics, traces, and logs, promoting interoperability and reducing vendor lock-in.
- **Wide adoption**: Supported by major cloud providers and observability vendors, ensuring compatibility and flexibility.
- **Cloud-native focus**: Designed for modern cloud-native architectures and Kubernetes environments.

Implementing these tools and techniques will require careful planning and configuration to address your specific use cases and needs. However, when properly implemented, they provide comprehensive observability of your systems across multiple clusters, regions, and cloud providers, giving you the high-level, global view of your infrastructure and applications' health and performance that you need.

Implementing observability

Implementing comprehensive observability in a multi-cloud Kubernetes environment involves several stages. It is not only about selecting the right tools but also about setting them up in a way that ensures seamless operation across different platforms, clusters, and regions. Here is a step-by-step guide on how to do it:

1. **Set your observability goals**: Before diving into the implementation, understand what you aim to achieve. Are you looking to troubleshoot faster, improve performance, or ensure service availability? Knowing your goals will guide your tool selection and how you configure them.

2. **Select your tools**: Choose your observability tools based on your goals, current stack, and team expertise. The tools should cover the three pillars of observability: logs (for example, ELK stack or Fluentd), metrics (such as, Prometheus and Grafana), and traces (like Jaeger or Zipkin). Moreover, consider a service mesh like Istio or Linkerd for automatic telemetry, and OpenTelemetry for a standardized approach to data collection.

3. **Design your observability architecture**: Now, design how these tools will interact with each other and with your applications. Alongside of your stakeholder requirements, consider how logs, metrics, and traces will be collected, stored, and analyzed. Think about the flow of data, data retention policies, and privacy considerations. Ensure your architecture supports multi-cloud environments, meaning that data from different clouds can be collected and viewed together.

4. **Deploy and configure your tools**: Possibly using the same GitOps method described previously install the selected tools in your clusters, taking into account the specifics of each cloud provider. Configure these tools according to your needs: set up log levels, metrics to collect, traces to record, and so on.

5. **Establish alerting rules**: Once data collection is set, define alerting rules in your monitoring tool using stakeholder requirements and industry best practices as your guide. These alerts should be actionable and tied to the SLIs/SLOs/SLAs of your services. They should notify your team about any abnormal conditions in real-time.

6. **Create visualizations**: Use tools like Grafana or Kibana to create dashboards that offer a high-level view of your systems' health and performance. Dashboards should be intuitive and give insights at a glance. Dashboards should be created for specific audiences. The customer service manager will not find anything helpful with a dashboard about pod health for example. In contrast the DevOps or SRE team would not find value in a sales dashboard.

7. **Continually review and update your setup**: Observability is not a set-and-forget operation. As your system evolves, so should your observability setup. Regularly review and update your tools, configurations, alerting rules, and visualizations to ensure they continue meeting your needs.

Implementing comprehensive observability in a multi-cloud Kubernetes environment can be challenging, but it is essential for maintaining system health and performance. With the right strategy, tools, and techniques, you can achieve a high-level, global view of your systems, clusters, infrastructure, and applications, enabling you to detect and resolve issues faster and make more informed decisions.

Observability with service mesh

A service mesh is a dedicated infrastructure layer that manages service-to-service communication in a transparent and language-agnostic way. It has emerged as a critical component in distributed systems like multi-cloud Kubernetes environments, primarily because of its inherent capabilities to enhance observability.

A service mesh like Istio, coupled with Envoy as the sidecar proxy, inherently provides deep observability features. These features include automatic telemetry data collection, real-time monitoring, tracing, and visualization. Let us delve into how a service mesh helps enhance observability:

- **Automatic telemetry data collection**: Service mesh architecture follows the sidecar pattern, where each service in the mesh has a proxy attached to it. This proxy intercepts all incoming and outgoing traffic, providing a wealth of information that can be used for observability. It can automatically generate detailed telemetry for all service communications, including error rates, latencies, and request volumes. If a new workload or service is launched, it automatically gets the same observability as all of its peers, without manual steps which could potentially be forgotten or skipped.

- **Real-time monitoring**: A service mesh continually monitors the network, collecting real-time metrics from each service instance in the mesh. It can offer insights into service behavior, network issues, performance bottlenecks, and dependencies among services, enabling you to detect and troubleshoot issues quickly. For example, directly after a code push a flood of HTTP 500 errors appear, with real time monitoring you are able to detect the issue and rollback to the last know good version reducing your **Mean Time to Recovery (MTTR)**.

- **Distributed tracing**: Service meshes often come with integrated tracing tools like Jaeger or Zipkin. These tools provide the capability to follow a request as it travels through various services in the mesh. Tracing helps to identify latency issues and understand the interaction between different microservices. For example, Istio supports OpenTracing and Envoy has native Zipkin support.

- **Visualization**: Service meshes often integrate with visualization tools like Kiali that allow you to visualize your mesh topology, inspect traffic flow between services, and monitor the health of different services. Some examples of valuable visualization include, topology maps of nodes and connections, traffic animations, custom filters and views and service details when clicking on a node.

- **Interoperability with observability tools**: Service meshes like Istio can integrate with popular observability tools like Prometheus and Grafana, enhancing their capabilities and offering a comprehensive monitoring solution. Prometheus can scrape metrics from Istio, and Grafana can visualize them, creating a powerful observability pipeline.

In a multi-cloud Kubernetes context, a service mesh can aggregate data from clusters running in different cloud environments, providing a unified, high-level view of the health and performance of your entire system. This capability is crucial in maintaining consistent observability practices across varying cloud environments and ensuring the global health of your systems.

In conclusion, incorporating a service mesh into your Kubernetes environment can significantly enhance your observability efforts. It provides an automated, in-depth, and real-time view of your system, helping you maintain high performance and availability.

Best practices for comprehensive observability

Observability is a key requirement for ensuring the health and performance of multi-region or multi-cloud Kubernetes environments. To make the most of your observability efforts, here are some best practices to follow.

Use standardized and structured logging

Ensure that all your services follow a standardized logging format. Structured logs, preferably in a machine-readable format like JSON, are easier to process, filter, and analyze. JSON offers machine readability, interoperability, flexibility, human readability, storage and transmission efficiency, is an open standard and is well supported in tools and libraries. This makes it the defacto standard for standardized and structured logging. This standardization will simplify the process of consolidating logs from multiple sources and enable you to extract meaningful insights efficiently.

Instrumentation

Ensure that your applications and infrastructure components are instrumented to emit metrics and traces. An example is tuning exporters with Prometheus or including OTEL libraries in your product code. It is essential to capture data on request rates, error rates, and latency as they provide valuable insights into your system's performance.

Leverage distributed tracing

In complex, distributed systems, it is crucial to understand how requests traverse through different services. Distributed tracing allows you to follow a request's journey, helping you identify bottlenecks and performance issues in your system. As an example, imagine an online store with a service mesh:

- **User places an order**: A request enters the mesh, triggering a chain of actions across multiple services:

- o **Product catalog**: Retrieves product details.
- o **Inventory**: Checks stock availability.
- o **Payment processing**: Handles payment authorization.
- o **Order fulfillment**: Triggers shipping process.

Distributed tracing follows the request's journey:

- **Each service generates spans**: As the request passes through each service, the service mesh creates a span, containing:
  - o Start and end timestamps
  - o Duration of the operation
  - o Metadata (for example, service name, request type, errors)
- **Spans are linked together**: The spans are connected, forming a trace representing the request's full path through the system.
- **Traces are collected and visualized**: The service mesh sends traces to a tracing backend (like Jaeger or Zipkin), which analyzes and visualizes them.

Set up efficient alerting

Observability is not just about collecting data; it is also about understanding this data and responding to it. Ensure that you have an effective alerting system in place that notifies you about potential issues before they escalate. Use alerts sparingly and wisely to avoid alert fatigue.

Correlation of data

It is crucial to be able to correlate events across logs, metrics, and traces to get a holistic view of your system. This practice helps in faster and accurate debugging. Using the online store example above, imagine a spike in error rates:

Here is a simple example of how data correlation works in a service mesh to achieve a holistic view and faster debugging. Imagine a spike in error rates in your online store:

- **Observing metrics:**
  - o Your monitoring dashboard shows a sudden increase in 5x errors in the checkout service.
  - o Metrics reveal high response times and increased latency.
- **Diving into logs:**
  - o You filter logs for the checkout service around the time of errors.
  - o You discover database connection timeouts in the logs, suggesting a potential root cause.

- **Tracing the journey:**
  - You initiate a distributed trace for a failed checkout request.
  - The trace visualizes the request's path, revealing a delayed database call within the checkout service.
- **Correlating insights:**
  - You correlate the metric spike, log messages, and trace evidence to pinpoint the database connection issue as the primary cause of errors.
- **Taking action:**
  - You investigate database performance, network connectivity, or query optimization to address the root cause.

Embrace service mesh

A service mesh can significantly enhance observability by automatically collecting detailed telemetry, providing real-time monitoring, and allowing effective visualization. Consider using a service mesh in your Kubernetes environment for improved observability.

Continual review and improvement

Observability is not a one-time setup and it requires continuous review and improvement. Regularly review your metrics, logs, and traces to ensure they are providing the required insights. Refine your observability practices based on the changing needs of your environment.

Use open standards and open-source software

This approach avoids vendor lock-in, promotes interoperability, and ensures a broad community support.

Ensure security

Security is paramount when implementing observability, especially in multi-cloud environments. Use encryption, manage access control, and ensure compliance with regulations to secure your observability data.

Observability is a journey, not a destination. It needs a culture of learning and improvement. By following these best practices, you will be better equipped to understand the health and performance of your systems, infrastructure, and applications in a multi-cloud Kubernetes context.

Challenges in implementing observability

Implementing comprehensive observability across a multi-cloud, multi-region Kubernetes infrastructure can be a complex endeavor. Here are some common challenges and ways to overcome them.

Volume of data

Multi-cloud Kubernetes deployments often generate massive amounts of observability data (logs, metrics, traces). This scale can make data management, storage, and analysis challenging.

Solution: Use tools that can efficiently handle large volumes of data. Consider using data reduction techniques such as sampling, aggregation, and compression. Additionally, apply retention policies to discard old data that is not needed.

Diversity of data sources

With the myriad components involved in a multi-cloud Kubernetes infrastructure, the data comes from numerous, diverse sources. This diversity makes it difficult to correlate data and gain meaningful insights.

Solution: Implement standardized logging formats across your systems along with well-defined tagging to help identify issues and aid in searches. Use a unified observability platform that can ingest data from various sources, normalize it, and present it in a coherent, unified view.

Complexity of systems

With microservices architecture, containers, and multiple cloud providers, the complexity of the system can be daunting. This complexity can make it hard to understand the interdependencies and track problems.

Solution: Use service meshes, which can help manage the complexity of microservices architectures by providing a uniform way to connect, manage, and secure microservices. Tools like distributed tracing also help to understand inter-service communications.

Alert fatigue

Too many insignificant alerts can lead to alert fatigue, causing teams to overlook critical alerts. Do you pay attention to car alarms anymore? Likely no, because they go off constantly without good reason, so you ignore them. The same idea applies here.

Solution: Establish alerting policies that only trigger alerts for critical issues. Use techniques like anomaly detection to identify unusual patterns and reduce the noise in your alerts.

Security and compliance

Especially in multi-cloud environments, ensuring the security and compliance of your observability data can be a challenge. Without legislation or a compliance framework in place it can be difficult to prioritize these.

Solution: Apply security best practices, including encryption, access controls, and auditing. Moreover, familiarize yourself with the regulations applicable to your industry and ensure that your observability practices comply with them.

Skills and knowledge gap

Observability requires a different mindset and a specific set of skills, which your team might not possess initially. Junior or new team members often need help closing this gap.

Solution: Invest in training your team. Encourage a culture of learning and continuous improvement.

Cost

The tools and infrastructure required for comprehensive observability can be expensive, especially when dealing with large volumes of data across multiple cloud platforms.

Solution: Consider open-source tools, which can offer robust observability solutions at a fraction of the cost. Moreover, monitor your usage and optimize to reduce costs.

By understanding and addressing these challenges, you can enhance your organization's ability to implement effective and comprehensive observability for your multi-region, multi-cloud Kubernetes environments.

Case studies: Observability in action

Observability can sometimes feel abstract without concrete examples, so let us explore a couple of case studies that highlight its value and impact in a multi-cloud, multi-region Kubernetes environment.

E-commerce giant embraces observability for Black Friday

An e-commerce giant gearing up for Black Friday, one of the busiest shopping days of the year, wanted to ensure their systems could handle the anticipated traffic surge. Their infrastructure spanned multiple regions and clouds to ensure high availability and redundancy.

To maintain control over this complex setup, they adopted comprehensive observability practices. They collected logs, metrics, and traces from each service and used service mesh technology to standardize and simplify their collection across regions and cloud providers.

When Black Friday arrived, they detected an unusual latency spike in one of their payment microservices. Thanks to their comprehensive observability, they quickly traced the issue back to a recent deployment, rolled it back, and resolved the problem within minutes. Without the visibility provided by their observability practices, this process could have taken hours and resulted in significant revenue loss.

Streaming service avoids outage during global event

A popular streaming service was preparing to stream a highly-anticipated global sports event. They ran their services on a multi-cloud, multi-region Kubernetes setup to deliver a seamless streaming experience to millions of viewers worldwide.

Using observability tools, they were able to monitor the health and performance of their systems in real-time. They set up dashboards that provided global views of their system, displaying key metrics like error rates, request latencies, CPU utilization, and more.

As the event started and traffic began to surge, their observability system detected a growing number of errors from a specific region. Thanks to their comprehensive observability setup, they quickly pinpointed the issue to a misconfigured load balancer in that region. They rectified the misconfiguration promptly, thereby avoiding a potential service disruption.

These case studies highlight the importance of comprehensive observability. It not only helps detect issues quickly but also aids in resolving them promptly, ensuring high availability and excellent user experience, especially crucial in multi-cloud, multi-region Kubernetes environments.

Conclusion

Congratulations! You have made significant progress in your learning journey and have now completed this chapter. In this chapter, we have dived deep into the world of comprehensive observability within the context of multi-region and multi-cloud Kubernetes environments. We started with an overview of observability, understanding its importance and relevance in today's complex system landscapes.

We broke down the three pillars of observability: logs, metrics, and traces, and discussed how each contributes to a holistic view of our system's health and performance. We then explored various tools and techniques that help achieve comprehensive observability in such distributed setups.

We examined how service meshes like Istio and Envoy play a significant role in achieving observability, and also delved into the best practices for implementing observability in a multi-cloud Kubernetes environment. Finally, we went through some real-world case studies, highlighting how observability can save the day in complex, high-stakes situations.

Your dedication and effort to learn these concepts will undoubtedly pay off in managing and operating robust, performant, and resilient systems in multi-cloud Kubernetes environments.

As we move on to the next chapter, we will be venturing into another crucial aspect of managing Kubernetes environments, that is, *Securing Your Clusters*. Here, we will talk about how to safeguard your clusters against potential threats, maintain data privacy, and ensure compliance with regulatory standards. Exciting and important content awaits you, so stay tuned!

Join our book's Discord space

Join the book's Discord Workspace for Latest updates, Offers, Tech happenings around the world, New Release and Sessions with the Authors:

https://discord.bpbonline.com

CHAPTER 12
Securing Your Clusters

Introduction

Welcome to this chapter, where we embark on the essential journey of securing our Kubernetes clusters. Security is a fundamental aspect of managing any digital system, and Kubernetes is no exception. As we progress into increasingly complex multi-cloud environments, the importance of robust, comprehensive security practices grows exponentially. In this chapter, we will explore securing our clusters both at the infrastructure level and the network level, focusing on protecting externally generated client requests as well as intra-cluster, app-to-app communications.

Structure

Here is an overview of the topics we will cover:
- Introduction to Kubernetes security
- Four Cs of cloud-native security
- Securing your infrastructure: Best practices
- Network security in Kubernetes
- Securing externally generated client requests
- Securing intra-cluster app-to-app communication

- Tools and techniques for Kubernetes security
- Case studies: Security practices in action
- Best practices for implementing security
- Challenges and solutions in Kubernetes security
- Security in the context of service mesh

Objectives

By the end of this chapter, you will have a solid understanding of various facets of Kubernetes security, starting with an understanding of the *four Cs*, that is, Cloud, Clusters, Containers, and Code, which provide a structured approach to security in Kubernetes. You will learn the best practices for securing your infrastructure and understand how to implement network security effectively. We will also introduce you to the tools and techniques commonly used for enhancing security in Kubernetes clusters. Finally, you will learn from real-world case studies, showcasing the implementation of these security practices and how they stand up to challenges. By understanding these concepts, you will be better equipped to create and maintain secure, reliable, and resilient multi-cloud Kubernetes environments. Your journey towards mastering Kubernetes security starts here!

Introduction to Kubernetes security

Security is a critical aspect of any software system, but it is particularly vital when dealing with a container orchestration system such as Kubernetes, especially in multi-cloud environments. Securing a Kubernetes environment involves protecting and monitoring multiple layers, from the infrastructure underlying your clusters to the network traffic flowing within and outside of them.

As our systems grow and extend over multiple clouds, the task of ensuring a robust and comprehensive security model becomes more complex but also more necessary. A key aspect of achieving this comprehensive security is understanding and utilizing a service mesh like Istio, which is integrated with Envoy. Istio and Envoy provide a powerful, extensible suite of security features that can significantly bolster the security posture of your Kubernetes clusters.

Istio's security capabilities extend across authentication, authorization, and network encryption, providing a secure communication channel for both your service-to-service and end-user-to-service transactions. It enables features such as automatic **Mutual TLS (mTLS)** encryption, fine-grained access control, and strong identity assertions. These tools facilitate secure, seamless communication within your cluster and provide a foundation for comprehensive observability and intelligent routing.

On the infrastructure side, Kubernetes provides its own set of security features that can protect your clusters. These include **Role-Based Access Control (RBAC)**, Network Policies, Pod Security Policies, among others. Understanding and properly implementing these features is critical for safeguarding your infrastructure.

But remember, Kubernetes security is not just about using the right tools, it also involves following best practices, understanding the potential threats, and continuously monitoring your environment for any signs of breach or malfunction. That is what we are going to delve into in the following sections of this chapter.

Four Cs of cloud-native security

In order to secure your multi-cloud Kubernetes clusters effectively, it is crucial to consider the **Four Cs** of cloud-native security: **Code, Container, Cluster, and Cloud.** This model offers a comprehensive approach to security, addressing it at each layer of a cloud-native stack. Here is a closer look at each C:

Figure 12.1: The Four Cs

Code

The **Code** layer of the four Cs of cloud-native security refers to the application code itself. This layer is responsible for implementing the security controls that protect the systems and networks involved in multi-cluster Kubernetes deployments across multiple cloud providers. Refer to the previous chapter on GitOps for more details here.

Some of the key security considerations for the Code layer include the following:

- **Secure coding practices**: The application code should be written in a secure manner, some examples of this are input validation, such as checking data types and ranges, sanitizing special characters, and rejecting invalid input. Also output encoding, such as encoding user-supplied data in output, using appropriate encoding functions and context-aware encoding. Also, exception handling such as implementing robust exception handling which catches exceptions gracefully to prevent application crashes, avoid revealing sensitive details in exception messages, providing generic error messages and logging exceptions securely.

- **Vulnerability scanning**: The application code should be scanned for vulnerabilities on a regular basis. This can be done using static code analysis tools or dynamic analysis tools. You can enable the use of static code analysis tools which analyze code without execution and focus on structural vulnerabilities. Some products include SonarQube, Checkmarx, Veracode, Fortify and Coverity. For dynamic analysis tools which analyize code during execution and focus on runtime behaviors, have a look at OWASP ZAP, Burp Suite, Acunetix, Netsparker and AppScan.
- **Dependency management**: The application code should depend on only trusted and secure dependencies. This can be done by using a dependency management tool that scans for known vulnerabilities in dependencies. Managing dependencies in a private repository like JFrog Artifactory can ensure the use of vetted libraries, binaries, or other dependencies.
- **Encryption**: The application code should encrypt sensitive data, such as passwords, API keys, and credit card numbers. This should be for data at rest and in transit. Refer to the chapter on service mesh for more details.
- **Access control:** The application code should implement access control mechanisms to restrict access to sensitive data and resources.

By following these security considerations, the Code layer can help to protect the systems and networks involved in multi-cluster Kubernetes deployments across multiple cloud providers.

Here are some additional details about how the Code layer contributes to overall security:

- **Restricting exposed endpoints, ports, and services**: The Code layer can be used to restrict the exposure of endpoints, ports, and services to only those that are necessary. This helps to reduce the attack surface and makes it more difficult for attackers to gain access to the system.
- **Protecting communication between services**: The Code layer can be used to protect communication between services using TLS encryption. This helps to ensure that sensitive data is transmitted securely and cannot be intercepted by attackers.
- **Validating input and output**: The Code layer can be used to validate input and output data to prevent attackers from injecting malicious code or data into the system.
- **Handling exceptions securely**: The Code layer can be used to handle exceptions securely to prevent attackers from exploiting vulnerabilities in the application code.

By implementing these security controls in the Code layer, organizations can help to protect their multi-cluster Kubernetes deployments across multiple cloud providers from a variety of attacks.

Container

The **Container** layer of the Four Cs of cloud-native security refers to the container images that are used to deploy applications in Kubernetes. This layer is responsible for ensuring that the containers are secure and that they do not contain any vulnerabilities that could be exploited by attackers.

Some of the key security considerations for the Container layer include the following:

- **Image scanning**: Container images should be scanned for known vulnerabilities on a regular basis. This can be done using static code analysis tools or dynamic analysis tools.
- **Signing images**: Container images can be signed to verify their authenticity and integrity. This helps to prevent attackers from tampering with the images or injecting malicious code.
- **Least privilege**: Containers should be run with the least number of privileges necessary to perform their tasks. This helps to reduce the attack surface and makes it more difficult for attackers to gain control of the container.
- **Network isolation**: Containers should be isolated from each other and from the host machine. This helps to prevent attackers from spreading malware or accessing sensitive data from other containers.
- **Secrets management**: Sensitive data, such as passwords and API keys, should be stored in a secure location and not in the container image itself. This helps to prevent attackers from accessing the data if they gain control of the container.

By following these security considerations, the Container layer can help to protect the systems and networks involved in multi-cluster Kubernetes deployments across multiple cloud providers.

Here are some additional details about how the Container layer contributes to overall security:

- **Restricting access to container registries**: Access to container registries should be restricted to only authorized users. This helps to prevent attackers from downloading malicious container images.
- **Using secure container runtimes**: Secure container runtimes provide additional security features, such as sandboxing and security isolation. This helps to protect the host machine from malicious containers.

- **Monitoring container activity**: Container activity should be monitored to detect suspicious behavior. This helps to identify and respond to attacks quickly. Refer back to previous chapters on observability for more details on this.

By implementing these security controls in the Container layer, organizations can help to protect their multi-cluster Kubernetes deployments across multiple cloud providers from a variety of attacks.

Cluster

The **Cluster** layer of the four Cs of cloud-native security refers to the Kubernetes cluster itself. This layer is responsible for securing the cluster and its components, such as nodes, pods, and services.

Some of the key security considerations for the Cluster layer include the following:

- **Network security**: The cluster network should be secured to prevent unauthorized access to pods and services. This can be done using network policies, firewalls, and VPNs.
- **Host security**: The host machines that make up the cluster should be secured to prevent unauthorized access and malicious activity. This can be done using security hardening techniques, such as patching, antivirus software, and intrusion detection systems.
- **Pod security**: Pods should be secured to prevent unauthorized access and malicious activity. This can be done using pod security policies, RBAC, and network policies.
- **Service security**: Services should be secured to prevent unauthorized access and malicious activity. This can be done using service accounts, RBAC, and network policies.
- **Logging and monitoring**: The cluster should be logged and monitored to detect suspicious activity. This helps to identify and respond to attacks quickly.

By following these security considerations, the Cluster layer can help to protect the systems and networks involved in multi-cluster Kubernetes deployments across multiple cloud providers.

Here are some additional details about how the Cluster layer contributes to overall security:

- **Using a secure Kubernetes distribution**: There are a number of secure Kubernetes distributions available, such as **Google Kubernetes Engine (GKE)**, **Azure Kubernetes Service (AKS)**, and **Amazon Elastic Kubernetes Service (EKS)**. These distributions come with several security features pre-configured, which can help to reduce the overall security risk.
- **Using a centralized configuration management system**: A centralized configuration management system can be used to manage the security configuration of the

cluster. This helps to ensure that the security configuration is consistent across all nodes and pods.

- **Using a secure cluster logging and monitoring system**: A secure cluster logging and monitoring system can be used to detect suspicious activity. This helps to identify and respond to attacks quickly.

By implementing these security controls in the Cluster layer, organizations can help to protect their multi-cluster Kubernetes deployments across multiple cloud providers from a variety of attacks.

Cloud

The **Cloud** layer of the four Cs of cloud-native security refers to the cloud infrastructure that hosts the Kubernetes cluster. This layer is responsible for securing cloud resources, such as virtual machines, storage, and networking.

Some of the key security considerations for the Cloud layer include the following:

- **Identity and Access Management**: Identity and Access Management (IAM) controls should be used to restrict access to cloud resources to only authorized users. This helps to prevent unauthorized access to sensitive data and systems.
- **Encryption**: Cloud resources should be encrypted to protect sensitive data from unauthorized access. This can be done using a variety of encryption techniques, such as TLS and SSL.
- **Logging and monitoring**: Cloud activity should be logged and monitored to detect suspicious activity. This helps to identify and respond to attacks quickly.
- **Security best practices**: The cloud provider's security best practices should be followed to help ensure that the cloud environment is secure. This includes things like patching systems regularly, using strong passwords, and implementing multi-factor authentication.

By following these security considerations, the Cloud layer can help to protect the systems and networks involved in multi-cluster Kubernetes deployments across multiple cloud providers.

Here are some additional details about how the Cloud layer contributes to overall security:

- **Using a secure cloud provider**: There are a number of secure cloud providers available, such as GCP and AWS. These providers offer a variety of security features, such as IAM, encryption, and logging and monitoring.
- **Using a secure cloud configuration**: The cloud configuration should be secure to prevent unauthorized access to cloud resources. This can be done by using a secure cloud provider and following the cloud provider's security best practices. Keep in mind this will vary between cloud providers so read the associated documentation carefully.

- **Using a secure cloud logging and monitoring system**: A secure cloud logging and monitoring system can be used to detect suspicious activity. This helps to identify and respond to attacks quickly.

By implementing these security controls in the Cloud layer, organizations can help to protect their multi-cluster Kubernetes deployments across multiple cloud providers from a variety of attacks.

By understanding and properly securing each of these four layers, you can create a robust and comprehensive security posture for your multi-cloud Kubernetes environment. Remember, security in a cloud-native world is not just about securing individual components but ensuring an end-to-end security approach.

Securing your infrastructure: Best practices

Securing your infrastructure in a multi-cloud Kubernetes environment involves several best practices. These practices are essential to safeguard your resources and data across multiple cloud providers while using Istio and Envoy for service mesh. Let us look at these best practices:

- **Infrastructure as Code (IaC)**: Using IaC tools such as Terraform or CloudFormation can help in ensuring that your infrastructure is consistently deployed with the correct configurations. This also means that infrastructure configuration can be version-controlled and audited for changes, adding an additional layer of security and accountability.
- **Least privilege access**: Always follow the principle of least privilege. Only give necessary permissions to services and users, limiting their potential access. This includes access to Kubernetes cluster-level resources, network access, and cloud resource permissions.
- **Secure network configuration**: Implement secure network policies in your Kubernetes clusters. With Istio, enforce strict mTLS for service-to-service communication to ensure traffic is secure and authenticated. On the cloud provider side, make sure to secure your VPCs, subnets, and other networking resources.
- **Encrypt sensitive data**: Always encrypt sensitive data at rest and in transit. Use **Key Management Systems** (**KMS**) provided by cloud providers for key management and rotation.
- **Secure IAM policies**: Implement secure IAM policies across your cloud environments. This includes ensuring that IAM roles and policies assigned to your Kubernetes clusters and nodes are not overly permissive.
- **Regular security audits**: Regularly audit your infrastructure for any potential security risks. This includes auditing your cloud resources, Kubernetes clusters, and network traffic. Tools like AWS Security Hub, GCP's Security Command

Center, or Azure Security Center can help with these audits in a multi-cloud environment.
- **Patch management**: Ensure that your infrastructure components are regularly updated and patched. This includes your Kubernetes clusters, nodes, and also your Istio and Envoy components. Regular patching reduces the attack surface by mitigating known vulnerabilities.
- **Backup and disaster recovery**: Ensure that you have a robust backup and disaster recovery plan in place. This should include regular backups of your important data and a plan to restore services quickly in case of a disaster.

By implementing these best practices, you can significantly enhance the security posture of your infrastructure in a multi-cloud Kubernetes environment.

Network security in Kubernetes

Network security in a multi-cloud Kubernetes environment is crucial to ensuring the integrity and confidentiality of communication within and across clusters. By leveraging Istio and Envoy as a service mesh, you can implement a range of best practices to ensure network traffic is secure:

- **Enforce network policies**: Network policies in Kubernetes allow you to control the traffic flow at the IP address or port level (OSI layer 3 or 4). Network Policies are Kubernetes resources that control the traffic between pods and endpoints. These are enforced by the network plugin, so you must be using a networking solution which supports network policy.
- **Leverage Istio for service-level security**: Istio's powerful service mesh capabilities enable granular, application-level security controls. This includes automatic mTLS encryption to secure service-to-service communication, RBAC for services, and secure ingress and egress traffic controls.
- **Secure ingress and egress traffic**: Kubernetes Ingress and Egress controls, particularly when managed through Istio, allow you to manage and secure external access to services within a cluster, and regulate outbound traffic. You can define rules for routing, load balancing, and SSL/TLS termination, among other tasks.
- **mTLS for service-to-service communication**: mTLS is a key feature of Istio. It provides two-sided SSL communication between services, ensuring that traffic is secure and trusted in both directions. This is critical for protecting data in transit and verifying the identity of services.
- **Implement DDoS protection measures**: Depending on the capabilities of your cloud provider, consider implementing DDoS protection measures to secure your Kubernetes service endpoints against volumetric attacks.

- **Audit and monitor network activity**: Regularly audit and monitor network activity within your Kubernetes clusters and across your multi-cloud environment. This can help identify unusual activity that could indicate a security concern.
- **Regularly update and patch**: Ensure your Istio, Envoy, and Kubernetes components are regularly updated and patched to address any known security vulnerabilities.

With these practices in place, you can ensure a robust network security posture for your multi-cloud Kubernetes environments, protecting both your infrastructure and the applications running upon it.

Securing externally generated client requests

Externally generated client requests to your multi-cloud Kubernetes environment can potentially expose your infrastructure and applications to various forms of attacks. Securing these requests is paramount to maintaining the security of your Kubernetes clusters. Here is how you can accomplish this using Istio, Envoy, and other security best practices:

- **Use Istio's ingress gateway**: The ingress gateway in Istio, built on Envoy, provides a secure way to manage access to services within your cluster from external sources. It enables you to control routing, load balancing, and SSL/TLS termination. Importantly, it also allows for the implementation of security measures such as rate limiting and IP whitelisting/blacklisting.
- **Enable mTLS**: With Istio, you can enforce mTLS for all incoming traffic. mTLS ensures that both client and server authenticate each other, providing a higher level of trust and confidentiality. Istio's mTLS setup can be used to authenticate and secure communication between the external client and the ingress gateway.
- **Use authorization policies**: Authorization policies in Istio allow you to define rules that specify what actions are permitted on the services in your cluster. These can be used to control which external clients can access which services, based on the details of the client request.
- **Implement rate limiting**: With Istio and Envoy, you can implement rate limiting to protect your services from being overwhelmed by too many requests at once. This can help mitigate certain types of Denial-of-Service (DoS) attacks.
- **Secure your load balancer**: Depending on your cloud provider, you may have a load balancer in front of your Kubernetes cluster. Make sure that this load balancer is properly configured and secured. This can include measures like setting appropriate access controls, regularly reviewing and updating security groups, and ensuring that SSL/TLS termination is properly configured.
- **Utilize Web Application Firewalls (WAF)**: Consider using a **Web Application Firewall (WAF)** to protect your applications from common web exploits. WAFs can be particularly useful for securing APIs and web interfaces that are exposed to external clients.

- **Frequent monitoring and logging**: Monitor your ingress traffic and log activity regularly. This helps in identifying unusual patterns that could indicate a security concern. Tools like Prometheus and Fluentd can be used in conjunction with Istio and Kubernetes for effective monitoring and logging.

Securing externally generated client requests is a critical aspect of safeguarding your multi-cloud Kubernetes environment. Following these guidelines can help maintain the security of your clusters and mitigate the risk of potential attacks.

Securing intra-cluster app-to-app communication

In a multi-cloud Kubernetes environment, secure intra-cluster app-to-app communication is vital for maintaining the overall integrity of your system. Both Istio and Envoy play significant roles in this regard. Here are some strategies for securing this type of communication:

- **Use mTLS**: With Istio, you can enable mTLS for all inter-service communications within the cluster. This ensures that both the client service and the server service authenticate each other, providing a higher level of trust and confidentiality. It encrypts the communication channel and ensures only authorized services can communicate with each other.
- **Implement service-to-service authorization**: Istio's authorization policy provides a flexible model to control service-to-service access based on conditions such as service identity, request principals, and IP addresses. These policies can help prevent unauthorized access and limit the potential blast radius if a malicious actor gains access to one of your services.
- **Use network policies**: Kubernetes network policies provide another layer of security, controlling which services can communicate with each other within your cluster. This can be used in conjunction with Istio's policies to create a robust, layered defense.
- **Employ fine-grained traffic control**: With Istio's traffic management features, you can control how and where your service-to-service traffic flows. This can be particularly useful for managing complex deployments and ensuring high availability.
- **Enable secure ingress and egress control**: Istio's ingress and egress gateways help control incoming and outgoing traffic to your services. Egress control is particularly important in the context of service-to-service communication, as it can prevent services from reaching out to untrusted external services.
- **Monitor and log communications**: Consistent monitoring and logging of service-to-service communications can help you identify any potential security threats.

Envoy's access logs, combined with Istio's monitoring features, can provide valuable insights into your service interactions.

- **Regularly update and patch your software**: Ensure all your services, as well as your Istio and Envoy installations, are regularly updated and patched. This can help prevent security vulnerabilities that could be exploited for unauthorized app-to-app communication.

By following these best practices, you can enhance the security of your intra-cluster app-to-app communication, thereby strengthening the overall security posture of your multi-cloud Kubernetes deployments.

Tools and techniques for Kubernetes security

Security is crucial in any environment, and when it comes to multi-cloud Kubernetes clusters using the service mesh Istio and Envoy, several tools and techniques can aid in ensuring your clusters remain secure. Here, we will explore some of the essential tools and techniques:

- **Istio for service mesh security**: Istio provides robust security features for service-to-service communication in a cluster. It provides identity-based security features, allowing you to enable mTLS for secure communication. It also offers flexible access control to protect services from unauthorized access.
- **Envoy for network security**: Envoy, as a high-performance proxy, provides out-of-the-box security features like TLS termination, HTTP/2 and gRPC proxies, and access logging, which can be highly beneficial for securing network communication.
- **Kubernetes network policies**: Network policies in Kubernetes provide a way of controlling traffic between pods and clusters. They can be used to establish a baseline of isolation for pods and to restrict network access to and from your application.
- **Security contexts**: Kubernetes security contexts let you define privilege and access control settings for a pod or container, including control over the security features of the pod.
- **RBAC**: It is a method of regulating access to a computer or network resources based on the roles of individual users within your organization. Kubernetes supports RBAC natively, and it can be used to manage access to the Kubernetes API.
- **Secrets management**: Secrets management in Kubernetes helps to store and manage sensitive information, such as passwords, OAuth tokens, and ssh keys, securely.
- **Image scanners**: Tools like Clair or Trivy can be used to scan container images for known vulnerabilities. This ensures that the deployed containers are secure and free from any known threats.

- **Static code analysis**: Static analysis tools such as Checkov or Terrascan can be used to examine the **IaC** templates for security misconfigurations.
- **Audit logging**: Kubernetes audit logs provide a record of the sequence of activities that have affected the system by individual users, administrators or other components.

Implementing a combination of these tools and techniques can help enhance the security of your Kubernetes clusters, be it single, multi-region, or multi-cloud deployments. Remember, security is not a one-time activity, but a continuous process that evolves with your infrastructure and applications.

Case studies: Security practices in action

In this section, we will examine how different organizations have implemented security best practices to protect their multi-cloud Kubernetes environments, particularly using Istio, Envoy, and other relevant tools.

Case study 1: FinTech company implements zero trust networking

A financial technology company with a Kubernetes deployment spanning multiple cloud providers utilized Istio service mesh to enforce a zero-trust networking model. With Istio's mutual TLS, they ensured secure and encrypted service-to-service communication, significantly reducing the risk of man-in-the-middle attacks. The company also utilized Istio's policy enforcement features to create fine-grained access controls.

Case study 2: E-commerce firm secures service mesh with envoy

A global e-commerce organization used Envoy as the sidecar proxy in their service mesh to secure intra-cluster communication. They utilized Envoy's dynamic configuration, observability, and security features like built-in support for mTLS to achieve high-level security.

Case study 3: Healthcare provider secures patient data

A healthcare provider dealing with sensitive patient data used Kubernetes RBAC to manage access to the Kubernetes API, ensuring that only authorized personnel could access sensitive resources. They also used Kubernetes' Secrets management to handle sensitive data securely.

Case study 4: Media company handles high traffic securely

A media company with high external client requests used network policies in Kubernetes to control the flow of traffic in their multi-cloud Kubernetes environment. With a well-defined ingress and egress policy, they managed to secure externally generated client requests efficiently.

Case study 5: Software company streamlines compliance

A software company operating in a heavily regulated industry used Pod Security Policies and Security Contexts to enforce compliance at the container level. They also utilized image scanning tools to ensure that their deployed containers were free from any known vulnerabilities, streamlining their compliance process.

Each of these case studies highlights how a multi-layered security approach, combined with the right tools and practices, can create a robust security posture for Kubernetes environments across multiple clouds. It is important to note that there is no one-size-fits-all approach to security, and practices should be tailored to meet the specific needs and risks of your organization.

Best practices for implementing security

In the context of multi-cloud Kubernetes environments, using the Service Mesh Istio and Envoy, and considering both infrastructure and network traffic, there are several best practices you should consider when implementing security:

- **Follow the principle of least privilege**: Every component interacting with your Kubernetes system should have the minimum permissions necessary to carry out its function. This applies to both human users and applications.
- **Secure container images**: Always use trusted and secure container images. Regularly scan these images for vulnerabilities and use image signing to ensure their authenticity.
- **Secure network traffic with mTLS**: Istio and Envoy provide mutual TLS, which should be used to encrypt traffic between services. This secures communication within your cluster and makes it harder for unauthorized parties to intercept sensitive data.
- **Leverage Kubernetes network policies**: Network policies define how groups of pods are allowed to communicate with each other and other network endpoints. They are an essential tool for securing network traffic in Kubernetes.

- **Use role-based access control**: Kubernetes RBAC allows you to specify who can access the Kubernetes API and what permissions they have. It is a critical part of controlling access to your Kubernetes system.
- **Secure your Kubernetes API**: Limit network exposure for your Kubernetes API and use strong authentication methods. The API is a common attack vector, and securing it is essential.
- **Encrypt sensitive information**: Use Kubernetes Secrets to handle sensitive information like credentials. Secrets are encrypted and stored securely in the etcd datastore, making them much safer than plain text configuration options.
- **Implement a zero-trust network policy**: Never trust, always verify. Even if a request comes from within your network, verify its authenticity before granting access. Tools like Istio can help to enforce a zero-trust network model.
- **Keep your Kubernetes cluster up-to-date**: Regularly update your Kubernetes cluster to benefit from the latest security patches and updates.
- **Regularly audit and monitor**: Regular auditing of your cluster's activities helps to identify any anomalies and suspicious activities. Use proactive monitoring tools to provide real-time alerts and visualization of the overall health of your system.

Each of these best practices contributes to creating a secure Kubernetes environment, even when spanning multiple cloud providers. By implementing these practices, you can significantly improve the security of your multi-cloud Kubernetes clusters.

Challenges and solutions in Kubernetes security

Securing a multi-cloud Kubernetes environment presents several unique challenges. Let us explore these challenges along with solutions and best practices to mitigate them:

- **Complexity of multi-cloud environments**
  - **Challenge**: In a multi-cloud environment, maintaining consistency and managing different cloud providers' security practices can be complex and challenging.
  - **Solution**: Use **IaC** tools such as Terraform to maintain consistency across cloud providers. Additionally, a service mesh like Istio and Envoy can abstract away some of the complexities related to security and provide a unified layer of security across all clouds.
- **Securing network traffic**
  - **Challenge**: In a distributed and dynamic environment like Kubernetes, securing network traffic both within the cluster and from external sources is a major challenge.

- o **Solution**: Use a service mesh to enable mTLS for secure communication within the cluster. For securing external traffic, employ Kubernetes network policies, ingress controllers, and a robust firewall strategy.
- **Securing Kubernetes API server**
  - o **Challenge**: The Kubernetes API server is a common target for attacks. Securing it is a top priority, but doing so without restricting legitimate use can be a challenge.
  - o **Solution**: Implement strong authentication and authorization methods like RBAC and limit network exposure of your API server. Regularly audit API server logs to detect and respond to any suspicious activities.
- **Managing secrets**
  - o **Challenge**: Kubernetes secrets used for storing sensitive information like database credentials, tokens, and so on, are by default not encrypted at rest, which is a major security concern.
  - o **Solution**: Ensure that secrets are encrypted at rest using KMS providers or third-party solutions. Always use secrets instead of plain text configurations for sensitive information.
- **Keeping up with updates and vulnerabilities**
  - o **Challenge**: Kubernetes is regularly updated with new security patches and features. Keeping up with these updates across a multi-cloud environment is a challenge.
  - o **Solution**: Regularly update your Kubernetes clusters using automation wherever possible. Consider using managed Kubernetes services provided by cloud providers that handle updates and patching automatically.
- **Maintaining compliance**
  - o **Challenge**: In a multi-cloud environment, maintaining compliance with industry-specific security standards can be complicated.
  - o **Solution**: Use **Policy as code (PaC)** to enforce compliance automatically and consistently across all your Kubernetes clusters, irrespective of the underlying cloud provider.

By understanding these challenges and implementing the suggested solutions, you can greatly enhance the security posture of your multi-cloud Kubernetes clusters.

Security in the context of service mesh

In the realm of multi-cloud Kubernetes, a Service Mesh like Istio and Envoy plays a pivotal role in enhancing the security posture of your application infrastructure.

A service mesh essentially provides a dedicated infrastructure layer for facilitating service-to-service communication in a secure, reliable, and observable way. Let us delve deeper into the security aspects of a service mesh:

- **Secure service communication**: Service meshes like Istio and Envoy leverage mTLS to encrypt communication between services in the mesh. This ensures that data transmitted over the network is always secure and can only be deciphered by the intended recipient. In the context of a multi-cloud Kubernetes environment, this significantly enhances the security of intra-cluster communication across different cloud platforms.
- **Fine-grained traffic control**: Service meshes provide capabilities for fine-grained traffic control. You can configure detailed routing rules, retries, failovers, and fault injection. These features can be used to enforce security policies and to protect against certain types of network attacks.
- **Identity and access control**: With a service mesh, every service gets a unique identity. This identity is used to establish mTLS connections and to enforce policies. Service-to-service access control can be implemented based on these identities, enabling you to control which services can communicate with each other.
- **Network policy enforcement**: Service mesh supplements Kubernetes network policies by providing additional, more granular control. It can enforce policies at the application layer (L7), which Kubernetes network policies do not cater to.
- **Audit and compliance**: All the communication within the service mesh is tracked and can be logged for audit purposes. These logs can be used to detect any anomalies or breaches and can serve as a key element in meeting various compliance requirements.

While implementing a service mesh in a multi-cloud Kubernetes environment comes with its own set of challenges, such as potential performance overhead, complexity of configuration, and the need for careful management, it brings robust security benefits that can substantially fortify your cluster security.

Conclusion

In this chapter, we delved into the critical aspects of ensuring the security of multi-cloud Kubernetes environments, with a focus on infrastructure and network security. We explored the four Cs of cloud-native security, including Code, Container, Cluster, and Cloud, and their significance in achieving a robust security posture.

Throughout the chapter, we discussed best practices for securing the infrastructure, network, and client requests in a multi-cloud Kubernetes setup. We explored various security tools and techniques, along with real-world case studies that showcased security practices in action. Additionally, we highlighted the challenges faced in Kubernetes security and provided solutions to overcome them. Lastly, we explored the importance of security in the context of service mesh and its role in securing intra-cluster communication.

In the next chapter, we will dive into a practical case study where we will explore the deployment of highly available applications across multiple cloud Kubernetes clusters. Stay tuned to learn about the implementation details and best practices for achieving high availability in a multi-cloud environment.

Join our book's Discord space

Join the book's Discord Workspace for Latest updates, Offers, Tech happenings around the world, New Release and Sessions with the Authors:

https://discord.bpbonline.com

Chapter 13
Case Study: Deploying Highly Available Application on Multi-Cloud Kubernetes

Introduction

In this chapter, we will delve into the exciting world of deploying **Highly Available (HA)** applications on Kubernetes across multiple cloud environments. Our focus will be on **Amazon Web Services (AWS)**, **Elastic Kubernetes Service (EKS)** and **Google Cloud Platform (GCP)**, **Google Kubernetes Engine (GKE)**. By the end of this chapter, you will have a comprehensive understanding of how to design, deploy, and manage resilient applications that can withstand failures and ensure continuous availability.

Structure

Here is an overview of the topics we will cover:
- Introduction to HA applications
- Definition and importance of HA
- Distinction between availability, reliability, and scalability
- Unlocking HA with Kubernetes
- Designing for HA
- Deploying applications on AWS EKS
- Deploying applications on GCP GKE

- Implementing cross-cloud communication
- Monitoring and managing HA
- Best practices for HA application deployment
- Challenges and considerations
- Case study: Deploying a multi-cloud HA application

Objectives

By the end of this chapter, readers will have gained a comprehensive understanding of the principles, strategies, and practical techniques required to deploy highly available applications on Kubernetes across multiple cloud providers. Through the exploration of real-world case studies, best practices, and challenges, readers will be well-equipped to design and implement resilient solutions in a multi-cloud environment, ensuring uninterrupted services and optimal performance for their applications.

Introduction to HA applications

In today's era of digital transformation, where downtime can cost businesses thousands to millions of dollars per minute, ensuring the HA of applications becomes paramount. Whether you are deploying stateless web front-ends or stateful databases, ensuring that these services are continuously available, irrespective of any infrastructure or software failures, is non-negotiable.

Definition and importance of HA

HA pertains to systems designed to be robust and operational over long stretches of time, with minimal outages or downtimes. This often implies redundant components, failover capabilities, and rapid recovery mechanisms. Some common examples include offsite storage of medical or finance related records, multiple web servers in multiple regions or clouds, having a multi-master DB or having multiple read DB's.

The importance of HA is as follows:

- **User experience:** In our 24/7 connected world, users expect applications to be available around the clock. Downtime can significantly degrade user experience, leading to loss of customers and revenue.
- **Business continuity**: Many enterprises rely on their applications for day-to-day operations. HA ensures business processes are not interrupted.
- **Reputation**: Frequent downtimes can tarnish an organization's reputation, leading to a loss of trust among its user base.

Distinction between availability, reliability, and scalability

Let us now go over the differences and distinctions between availability, reliability and scalability:

- **Availability**: Availability refers to the system's operational performance and ability to operate without interruption. An available system is one that is in a state to perform its designated function whenever it is required. For example, a web server that is 100% availability and is up 24 hours a day 7 days a week.
- **Reliability**: Reliability is about the system's capability to function without failure. It is about the consistency and trustworthiness of the application. Reliable systems may encounter issues, but they can recover without affecting availability. For example, a web server that is reliably serves requests with HTTP 200 response and no HTTP 500 responses.
- **Scalability**: Scalability deals with a system's capability to grow and manage an increased demand effectively. It is not just about adding more resources, but about how the system handles that growth—whether it is horizontal (adding more machines) or vertical (adding more power to the same machine).

Unlocking HA with Kubernetes

One of the most important metrics that can be used to gauge the viability and robustness of a cloud-native application is its high availability. This includes its resilience to failure, its ability to adapt to changing demand, and continue to function as expected under duress. Next, we explore some key components that contribute to tunning highly available applications.

The Kubernetes HA fundamentals are explained as follows:

- **Self-healing**: Kubernetes continuously monitors nodes and pods. If a node or pod fails, it reschedules the pod onto another node, ensuring availability. If the replicaset count is above 1 then the deletion and recreation of the failing pod will allow the service to continue without interruption and heal itself in the process.
- **Automated rollouts and rollbacks**: Kubernetes progressively rolls out changes to your application or its configuration, ensuring system stability. If something goes wrong, it can revert the change. This is a larger topic and is not covered in this book.

Service mesh with Istio and Envoy

The benefits of using service mesh with Istio and Envoy are explained as follows:

- **Traffic management**: Using Istio with Envoy, you can control the traffic flow, do canary deployments, and ensure that only healthy services get traffic.
- **Security**: End-to-end authentication, authorization, and encryption enhance security without modifying the application.
- **Observability**: With a service mesh, you get insights into how services communicate, helping in debugging and optimization.

Kubernetes and multi-cloud

The benefits of using Kubernetes in a multi-cloud scenario are explained as follows:

- **Platform agnostic**: Kubernetes abstracts away cloud-specific details. This means you can run your Kubernetes cluster on AWS, GCP, Azure or even on-premises, providing flexibility and avoiding vendor lock-in. This is one of the most important benefits of running workloads in this manner!
- **Unified API and operations**: No matter where you run your applications, Kubernetes provides a unified API and set of operations, simplifying multi-cloud deployments and management.

In summary, HA is no longer a luxury but a necessity for modern applications. Kubernetes, combined with a service mesh like Istio and Envoy, provides a robust foundation for building HA applications that can run consistently and reliably across multiple cloud environments.

Refer to the following figure:

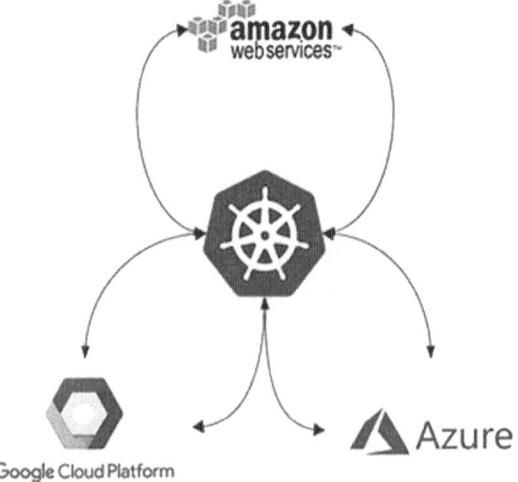

Figure 13.1: Multi-cloud Kubernetes

Designing for HA

Designing applications for HA in a cloud-native environment demands a keen understanding of various architectural and infrastructural considerations. The interplay of managed Kubernetes services like EKS and GKE, combined with the capabilities of Istio and Envoy as a service mesh, offers a vast array of tools and patterns to ensure HA.

Factors influencing design

Let us now go over the three factors influencing design.

Latency

Definition: The delay between a client's request and a server's response.

Considerations: When deploying across multi-cloud or globally, latency becomes a major concern. Ensuring services are geographically close to the users or other dependent services can mitigate this.

Istio & Envoy Role: Leveraging Istio's traffic routing rules, you can direct traffic based on request origins, ensuring data travels the shortest path or avoids congested networks.

Redundancy

Definition: The duplication of critical components to ensure system functionality even if part of the system fails.

Considerations: With managed Kubernetes, it is simpler to ensure redundant nodes, pods, or clusters. This is crucial for both EKS and GKE to ensure that if one service in one region or zone goes down, another can take its place.

Istio & Envoy Role: They allow for easy traffic splitting, enabling you to divert traffic to backup services or regions if primary ones fail.

Replication

Definition: Creating copies of data to ensure its availability and durability.

Considerations: Stateless applications can easily be replicated across zones or clusters. Stateful applications, however, require consistent data replication strategies.

Istio & Envoy Role: While these primarily deal with traffic, they ensure that replicated services (across regions/zones) receive their share of traffic without inconsistencies.

Stateful vs. stateless applications

Let us understand stateful and stateless applications in terms of considerations and design patterns.

Stateless applications

Definition: Applications that do not retain user session data between requests.

Considerations: Stateless applications are easier to scale since any request can be served by any instance. They fit naturally into the containerized world. See the previous chapter on Stateless application for more details.

Design patterns: Horizontal scaling, where new instances are created as demand increases, is straightforward.

Stateful applications

Definition: Applications that maintain session data or state between requests.

Considerations: They require persistent storage and a strategy to ensure data consistency and availability. See the previous chapter on Stateful application for more details.

Design patterns: Sticky sessions, where a user's all requests are directed to the same instance. Database replication (master-slave or master-master) ensures data consistency across instances.

Role of persistent storage and network design

The role of persistent storage and network design are as follows.

Persistent storage

Importance: For stateful applications, preserving data beyond the lifecycle of a pod is essential.

Solutions: Both AWS and GCP provide persistent storage solutions, like **Elastic Block Store (EBS)** for EKS and Persistent Disk for GKE. Integrating these with Kubernetes ensures that even if a pod dies, its data remains intact.

Dynamic provisioning: Through Kubernetes, it is possible to automatically provision storage as needed, ensuring elasticity.

Network design

Importance: Ensuring low latency, high throughput, and fault tolerance.

VPC & Subnets: Both AWS and GCP allow for the creation of **Virtual Private Clouds** (**VPC**) and subnets, ensuring isolation and security.

Load balancing: Crucial for distributing incoming traffic across services. Both EKS and GKE offer native load balancing solutions.

Istio and Envoy Role: They further enhance network design by providing sophisticated routing rules, security (mTLS), and traffic control.

In conclusion, designing for high availability in a Kubernetes environment is a multifaceted endeavor. It involves understanding the nature of your applications (stateful vs. stateless), ensuring data persistence, and creating a robust network design. Managed services like EKS and GKE, combined with Istio and Envoy, offer a robust toolkit to ensure that applications remain highly available, resilient, and scalable.

Deploying applications on AWS EKS

Amazon's **EKS** provides a managed Kubernetes service that makes it easier for users to run Kubernetes on AWS without managing the underlying infrastructure. The synergy of EKS, when combined with tools like Istio and Envoy, offers a dynamic platform to deploy both stateless and stateful applications. Here is a detailed look.

Key features and benefits

Let us discuss some key features and benefits in this section.

Key features

Some features are as follows:
- **Managed Kubernetes Control Plane**: AWS handles the maintenance, scalability, and high availability of the control plane for you.
- **Integration with AWS Services**: EKS is deeply integrated with services like VPC, IAM, and CloudWatch.
- **Hybrid deployments**: You can seamlessly integrate your on-premises deployments with AWS using EKS.
- **Modular and scalable**: Utilize AWS Fargate for serverless compute or EC2 for more control over the infrastructure.

Benefits

Some benefits are as follows:

- **Reliability and security**: EKS automatically manages the availability and scalability of the Kubernetes control plane nodes.
- **Consistent experience**: As EKS uses upstream Kubernetes, there is no variation in API behavior.
- **Deep monitoring**: Integration with CloudWatch and X-Ray provides a deep insight into the cluster's performance.

Steps for setting up an EKS cluster.

Follow the given sections to learn how to set up an EKS cluster.

EKS control plane creation

Follow the given steps:

1. **Set up AWS CLI**: Ensure AWS CLI is installed and configured with the necessary permissions.
2. **Create a VPC**: This will house your EKS cluster. Utilize the AWS Management Console or AWS CLI.
3. **Launch EKS**: Use the AWS Management Console to create a new EKS cluster within the VPC.
4. **Configure IAM roles**: Attach roles for worker nodes to communicate with the EKS master.

Node setup

Follow the given steps:

1. **Choose node infrastructure**: Decide between EC2 instances or Fargate for your worker nodes.
2. **Launch Worker Nodes**: Use AWS Management Console to launch the nodes in the earlier created VPC.
3. **Connect Nodes to EKS**: After launching, they will automatically connect to the EKS control plane.

Deploying an application using Helm and other EKS tools

To deploy an application using Helm and other EKS tools, follow the given sections.

Setting up Helm

Follow the given steps:

1. **Install Helm CLI**: Ensure you've Helm v3 (or latest) installed locally.
2. **Initialize Helm**: No need to install Tiller (used in Helm v2) with Helm v3.

Deploying using Helm

Follow the given steps:

1. **Add chart repositories**: Use `helm repo add` to add chart repositories.
2. **Search for charts**: Use `helm search repo` to find available charts.
3. **Install the chart**: For instance, to deploy a stateful application like MySQL, use `helm install [name] stable/mysql`.

EKS tools and utilities

Follow the given steps:

1. **eksctl**: A simple CLI utility for creating and managing clusters on EKS. It abstracts a lot of complexities.
2. **AWS Load Balancer Controller**: Allows you to manage AWS Load Balancers (ALB/NLB) directly via Kubernetes native service and ingress resources.

EKS with Istio and Envoy

Once the cluster is up, install Istio using Helm or Istio's custom installation methods. With Istio installed, you can leverage Envoy for powerful service mesh capabilities, providing resilient, secure, and dynamic networking for your applications.

In summary, AWS EKS provides a potent, scalable, and flexible platform for deploying applications. Whether you are deploying a simple stateless web application or a complex stateful database cluster, EKS, combined with tools like Helm, Istio, and Envoy, provides the capabilities you need to ensure high availability and smooth operation.

Deploying applications on GCP GKE

GKE is a powerful managed Kubernetes service that allows businesses and individuals to deploy, manage, and scale containerized applications using Google's infrastructure. Like EKS on AWS, GKE provides a seamless environment to run both stateless and stateful applications, further enhanced by the capabilities of Istio and Envoy as a service mesh. Here is an in-depth dive.

The distinguishing features for GCP GKE are as follows:

- **Autopilot**: This mode automates the cluster's infrastructure management tasks, from node provisioning to auto-upgrades.
- **Node pools**: Allows the grouping of nodes within a cluster, where each pool can have a different configuration.
- **Google-OAuth integration**: Simplifies access control by integrating with Google Cloud Identity.
- **Network policies**: Native support to control communication between pods.

Advantages

The advantages of GCP GKE are as follows:

- **Automated operations**: Automated repairs, upgrades, and scaling reduce the operational burden.
- **Security**: Provides end-to-end encryption, automatic OS upgrades, and private cluster options.
- **Global reach**: Integrated with Google's global load balancer and Cloud CDN.
- **Integrated developer tools:** Native integration with tools like Cloud Build, Cloud Source Repositories, and more.

Configuring a GKE cluster: Step-by-step guide

To configure a GKE cluster, follow the given sections.

Initial setup

Follow the given steps:

1. **Set up Google Cloud SDK**: Ensure gcloud CLI is installed and authenticated.
2. **Enable GKE API**: On Google Cloud Console, navigate to **Kubernetes Engine** and enable the API.

Cluster creation

Follow the given steps:

1. **Access Kubernetes Engine**: On the Console, go to Kubernetes Engine | Clusters.
2. **Click Create**. Choose between a Standard or Autopilot cluster, based on your needs.
3. **Configure cluster details**: Name your cluster, select a region or zone, and choose the Kubernetes version.
4. **Node configuration**: Select the machine type, disk size, and node image. For advanced configurations, you can modify node pools.
5. **Advanced Features**: Enable network policies, private clusters, or Cloud Monitoring/Logging based on your requirements.
6. **Click** Create: GCP will provision and set up the GKE cluster for you.

Application deployment using GCP-native tools

Let us learn about application deployment using GCP-native tools.

Container registry

Follow the given steps:

1. **Pushing to registry**: Build your Docker image and push it to Google Container Registry using gcloud docker -- push.
2. **Accessing images**: GKE can natively access these images, ensuring a smooth deployment process.

Deploying with cloud build and cloud source repositories

Follow the given steps:

1. **Source repository**: Push your application's source code to Cloud Source Repositories.
2. **Build with Cloud build**: Trigger an automated build using Cloud Build, which creates a container image and stores it in Container Registry.
3. **Deploy to GKE**: Use a `deployment.yaml` file that references the built image and deploy using `kubectl apply -f deployment.yaml.`

Cloud deployment manager

Follow the given steps:

1. **Templates**: Use Deployment Manager templates to define your application's resources.
2. **Deployment**: Deploy your application's infrastructure as code using the gcloud tool.

GKE with Istio and Envoy

After setting up the GKE cluster, you can enable Istio directly from the GKE interface or manually install it. With Istio in place, Envoy proxy will automatically be injected into your pods, establishing a powerful service mesh.

In essence, Google Kubernetes Engine offers a blend of advanced features and simplicity for deploying containerized applications. Its native integration with other GCP services makes the end-to-end development and deployment process more cohesive and efficient. With the added capabilities of Istio and Envoy, users can harness advanced traffic routing, telemetry, and security features seamlessly.

Implementing cross-cloud communication

In the era of cloud-native applications, many organizations are opting for multi-cloud deployments to tap into the unique features and capabilities of different cloud providers. However, setting up efficient cross-cloud communication, especially for containerized applications running on platforms like EKS and GKE, presents its own set of challenges. Using tools like Istio and Envoy can significantly simplify this process. Let us delve into the intricacies of implementing such a system.

Exploring challenges of cross-cloud communication

The challenges of cross-cloud communication are as follows.

- **Latency**: Different cloud providers might have data centers in varying locations, leading to latency when communicating between these centers.
- **Cost implications**: Data transfer between clouds is often billed, which can add a significant cost, especially for high-bandwidth applications.
- **Network configuration**: Each cloud provider has its own network setup and configurations, making uniformity a challenge. Ensure that you read the documentation for each to be aware of the differences.

- **Service discovery**: With applications spread across multiple clouds, keeping track of which service runs become intricate.
- **Security concerns**: Ensuring encrypted and authenticated communication between applications across different providers can be complex.
- **Compliance and regulations**: Data sovereignty laws might prevent certain data from being transferred between data centers in different regions or countries.

Configuring secure networking between AWS and GCP

Follow the given steps:

1. **VPN connectivity:** Establish a secure **Virtual Private Network (VPN)** connection between VPCs in AWS and VPCs in GCP. AWS offers VPN solutions that can connect with GCP's Cloud VPN.
2. **Direct connect and interconnect:** For a more stable and faster connection, consider using AWS Direct Connect and Google Cloud Interconnect. These provide direct physical connections between the two cloud providers.
3. **Uniform network policies:** Implement consistent network policies across both clouds. Istio can be pivotal here by providing consistent traffic controls across multi-cloud deployments.
4. **End-to-end encryption:** Ensure all data transferred between clouds is encrypted. Istio, with its in-built mTLS capability, can encrypt traffic between services seamlessly.

Demonstrating communication between applications across clouds

To know more about communication between applications across clouds, refer to the following sections.

Service entry and egress gateways

Follow these steps:

1. **Service entry**: Define a ServiceEntry in Istio to allow traffic to external services. This acts as a way to specify that a particular external service is accessible.
2. **Egress gateway**: Use Istio's Egress Gateway to manage traffic as it exits the mesh. This provides a security layer as you can apply policies to traffic exiting the service mesh.

Demonstration setup

Follow these steps:

1. **Deploy services**: Deploy a stateless application on EKS and another complementary service on GKE.
2. **Istio configuration**: On both EKS and GKE, configure Istio Service Entries to recognize the service on the other cloud.
3. **Initiate communication**: From the application on EKS, make a request to the service on GKE. With the correct Istio configuration, the service on EKS should be able to discover, communicate, and receive a response from the GKE service seamlessly.
4. **Monitor traffic**: Use tools like Kiali (which integrates with Istio) or cloud-native monitoring tools to visualize the traffic flow between services across clouds.

In summary, while cross-cloud communication comes with its set of challenges, leveraging the capabilities of service meshes like Istio, combined with cloud-native solutions, can help streamline the process. Not only does it allow for secure and efficient communication, but it also provides valuable insights into traffic patterns and service health across multiple cloud platforms.

Monitoring and managing HA

Ensuring HA is not just about the initial deployment and setup but involves continuous monitoring, alerting, and regular disaster recovery drills. As applications expand, so does their complexity. Therefore, having robust monitoring and management strategies in place is crucial for maintaining the HA promise. With platforms like EKS, GKE, and service meshes like Istio and Envoy, organizations have powerful tools at their disposal.

Importance of proactive monitoring in HA applications

The importance of proactive monitoring in HA applications are as follows:

- **Early detection**: Proactive monitoring helps in detecting issues before they escalate into more significant problems or outages. This allows for faster response times and minimizes service disruption.
- **Performance insights**: Continuous monitoring provides insights into how applications are performing, helping to identify areas of optimization.
- **Capacity planning**: By keeping an eye on resource utilization, organizations can anticipate when they will need to scale resources up or down.
- **Security**: Monitoring can also detect unusual patterns or breaches, which could indicate a security incident.

Tools: Prometheus, Grafana, and cloud-native solutions

Let us know more about tools now.

- **Prometheus:**
  - **Role**: A powerful open-source monitoring solution that integrates well with Kubernetes and service meshes.
  - **Features**: Real-time monitoring with a multi-dimensional data model. It can scrape metrics from configured targets at given intervals, evaluate rule expressions, and trigger alerts.
  - **Integration with Istio and Envoy**: Istio, combined with Envoy, emits metrics that can be ingested by Prometheus, providing deep insights into the mesh's traffic.
- **Grafana:**
  - **Role**: A leading open-source platform for visualization and analytics.
  - **Features**: Offers beautiful visualizations for time-series data. It can pull metrics from Prometheus and present them in a more readable and insightful manner.
  - **Dashboards**: Grafana dashboards for Istio provide out-of-the-box visualizations for metrics like request rate, error rate, and latency.
- **Cloud-native solutions:**
  - **AWS CloudWatch**: Offers insights into AWS resources, applications, and services that run on AWS and on-premises servers.
  - **Google Cloud** Monitoring: Provides visibility into the performance, uptime, and overall health of cloud-powered applications on GCP.

Alerting, Logging, and Disaster recovery strategies

Let us know more about these strategies.

Alerting

Tools like Prometheus can be configured to send alerts based on specific criteria. Integration with platforms like Alertmanager or PagerDuty can help in notifying the right teams.

Cloud-native solutions like CloudWatch Alarms or Google Cloud Alerting can also send notifications based on predefined conditions.

Logging

Centralized logging is crucial for debugging and understanding application behavior. Fluentd, Elasticsearch, and Kibana (often known as the EFK stack) are popular in the Kubernetes world.

AWS offers CloudWatch Logs, and GCP provides Cloud Logging for native logging solutions.

Disaster recovery strategies

The various disaster recovery strategies are as follows:

- **Backup and restore**: Regularly backup application data, configurations, and databases. Tools like Stash or Velero can assist in backup and restore procedures for Kubernetes.
- **Failover procedures**: Have automated processes in place to redirect traffic to healthy instances or regions should a part of the infrastructure fail.
- **Regular drills**: Conduct disaster recovery drills to ensure that the team knows the steps and that the backups are functional.
- **Documentation**: Maintain comprehensive and up-to-date documentation on disaster recovery procedures.

In essence, monitoring and managing high availability go hand in hand. It is not sufficient to have a highly available infrastructure if there are not adequate mechanisms to monitor its health and take swift action when needed. Using a combination of open-source and cloud-native tools can provide a holistic view of the system, ensuring that applications remain available, resilient, and performant.

Best practices for HA application deployment

Deploying applications with HA in a multi-cloud environment requires meticulous planning and adherence to best practices. With managed Kubernetes services like EKS and GKE combined with Istio and Envoy's capabilities, organizations can achieve a robust, resilient, and scalable infrastructure. Here is a deep dive into the core best practices for HA application deployment.

Ensuring redundancy at every layer

To reduce redundancy at every layer, refer to the following sections.

- **Infrastructure level:**
  - **Multi-zones and multi-regions:** Deploy clusters across multiple availability zones and regions. This ensures that if a particular zone or region experiences an outage, your application remains unaffected.
  - **Load balancers:** Use multi-zone or regional load balancers. AWS provides **Elastic Load Balancing (ELB)**, while GCP offers Global Load Balancers.
- **Application level:**
  - **Replica sets:** Always deploy applications using replica sets or deployments in Kubernetes. This ensures multiple replicas of your pods, so even if one pod fails, others can take over.
  - **StatefulSets for Stateful Workloads:** For stateful applications, use StatefulSets in Kubernetes. It ensures that even if a node fails, the stateful application will restart on another node with the same state.
- **Data level:**
  - **Database replication:** Employ master-slave replication or multi-master setups for databases. For instance, AWS RDS offers multi-AZ deployments, and Google Cloud SQL supports high-availability configurations.
  - **Distributed storage:** Use distributed storage solutions that automatically replicate data across nodes or zones.

Automated scaling

Let us learn more about automated scaling.

- **HPA:**
  - **Function:** HPA automatically scales the number of pods in a deployment or replica set based on observed metrics like CPU usage or custom metrics.
  - **Best practice:** Define clear metrics and thresholds for autoscaling. Monitor and adjust these as the application's requirements evolve.
- **Cluster autoscaling:**
  - **Function:** While HPA scales the pods, cluster autoscaler scales the actual nodes. If there are not enough resources to schedule a pod, the cluster autoscaler adds a new node.

o **Best practice**: Regularly review the minimum and maximum node counts. Overprovisioning can lead to unnecessary costs, while under provisioning can hinder scaling during traffic spikes.

Regular backup, testing, and update policies

Let us learn more about backup, testing and update policies.

- **Backup policies:**
  o **Frequency**: Depending on the application's criticality, decide on backup frequencies - daily, weekly, or even hourly for highly critical applications.
  o **Offsite backups**: Store backups in a different region or cloud provider. Tools like Velero can assist in Kubernetes backup and restore procedures.
- **Testing:**
  o **Chaos engineering**: Introduce failures deliberately in a controlled environment to test the system's resilience. Tools like Chaos Monkey or Litmus can be instrumental.
  o **Staging environment**: Always test new releases in a staging environment that mirrors the production setup as closely as possible.
- **Update policies:**
  o **Rolling updates**: Kubernetes supports rolling updates, ensuring zero downtime during application updates. It gradually replaces old versions with new ones, ensuring uninterrupted service.
  o **Canary deployments**: Deploy new versions to a subset of users to test before a full rollout. Service meshes like Istio simplify canary deployments.
  o **Security patches**: Regularly update the system and applications with security patches. Monitoring solutions can notify administrators about critical vulnerabilities.

In summary, while tools and platforms offer extensive features to ensure high availability, the onus is on organizations to implement them diligently and stick to best practices. Proper redundancy, proactive scaling, and regular maintenance are the pillars of a robust, highly available application deployment in a multi-cloud environment.

Challenges and considerations

Navigating the waters of multi-cloud Kubernetes deployments, especially with stateful and stateless applications and the use of service meshes like Istio and Envoy, comes with its unique set of challenges and considerations. While the promise of high availability, resilience, and scalability is tempting, there are hurdles that organizations must address proactively.

Data sovereignty and regulatory

Let us learn more about data sovereignty:

- **Location-specific regulations**: Various jurisdictions have regulations that dictate where and how data can be stored and transferred. For instance, the EU's **General Data Protection Regulation (GDPR)** has specific requirements about data storage and transfer.
- **Data residency**: Multi-cloud deployments mean that data can reside in multiple locations, possibly spanning various countries. Ensuring that data does not unintentionally cross borders is crucial.
- **Audit and compliance**: Regular audits might be required to ensure compliance with various regulations. The complexity increases when dealing with multiple cloud providers and geographies.
- **Encryption**: Ensure that data is encrypted both in transit and at rest. While cloud providers offer encryption tools, using them consistently across clouds is essential.

Cost management and optimization strategies

Let us learn more about cost management and optimization strategies:

- **Complex billing**: Multi-cloud means dealing with billing from multiple providers. Each provider has its own pricing model, leading to complexities in understanding and predicting costs.
- **Unused resources**: Resources that are not de-provisioned or are over-provisioned can lead to unnecessary expenses. Tools like AWS Trusted Advisor or Google Cloud's Cost Management tools can provide insights.
- **Traffic costs:** In a multi-cloud setup, data transferred between clouds might incur additional charges. Minimizing cross-cloud traffic is crucial.
- **Optimization tools**: Utilize cloud-native tools that help in cost prediction and optimization. Additionally, third-party tools like CloudHealth or Cloudability can give a holistic view across providers.

Vendor lock-in and interoperability challenges

Let us now learn about the various challenges.

- **Proprietary services**: While core Kubernetes might be consistent across clouds, cloud providers offer additional services that can be proprietary. Relying heavily on these can lead to vendor lock-in.

- **Service mesh complications**: While Istio and Envoy provide a level of abstraction, configurations might differ between cloud environments. Ensure configurations are as standardized as possible.
- **API inconsistencies**: Different cloud providers might have variations in their APIs, leading to challenges in automation and orchestration.
- **Migration considerations**: Always have a strategy for migrating services between providers. This includes data migration, reconfiguring networking, and understanding service differences.
- **Multi-cloud management platforms**: Consider using platforms like Anthos or Azure Arc, which aim to provide consistent management layers across different cloud providers.

In summary, while multi-cloud strategies bring diversification of risk and high availability, they also introduce complexities in terms of cost, compliance, and vendor dependencies. Organizations need to be well-prepared and strategic, considering both the benefits and challenges in their multi-cloud journey.

Case study: Deploying a multi-cloud HA application

In this final section of the chapter, we will apply the knowledge gained throughout the book to a comprehensive case study. Our goal is to deploy a highly available application on multi-cloud Kubernetes using the concepts, best practices, and techniques covered in previous chapters. This case study will provide readers with a real-world scenario that demonstrates the complete lifecycle of a multi-cloud deployment, from planning and architecture to implementation and monitoring.

Project overview

Our case study involves deploying a customer-facing e-commerce application that spans multiple cloud platforms for enhanced availability and scalability. The application's architecture will leverage Kubernetes to ensure high availability, fault tolerance, and efficient resource utilization.

Milestones

The various milestones are as follows.

Milestone 1: Planning and architecture

Refer to the following:

- Define the application's requirements, including performance, availability, and scalability.
- Choose the cloud providers: AWS and GCP.
- Design the multi-cloud architecture, incorporating load balancing, geographic distribution, and redundancy.

Milestone 2: Infrastructure setup

Refer to the following:

- Create AWS EKS and GCP GKE clusters with appropriate configurations.
- Implement networking and connectivity between the clusters using a service mesh (Istio and Envoy).

Milestone 3: Application deployment

Refer to the following:

- Containerize the e-commerce application components using Docker.
- Deploy the application to Kubernetes clusters on AWS EKS and GCP GKE.
- Utilize Kubernetes Deployments for scaling and self-healing of application pods.

Milestone 4: HA and failover

Refer to the following:

- Implement multi-cloud load balancing to distribute traffic across clusters.
- Set up database replication and synchronization for data consistency.
- Configure Kubernetes Ingress controllers and service mesh for seamless failover.

Milestone 5: Security and observability

Refer to the following:

- Apply security best practices, including **role-based access control (RBAC)** and network policies.
- Implement observability through comprehensive logging, monitoring, and tracing.

Milestone 6: Continuous deployment

Refer to the following:
- Set up GitOps workflow for managing application deployments and updates.
- Integrate CI/CD pipelines for automated testing and continuous deployment.

Milestone 7: Testing and validation

Refer to the following:
- Perform rigorous testing, including chaos testing and failure simulations.
- Validate the application's behavior during various failure scenarios.

Milestone 8: Performance optimization

Refer to the following:
- Optimize resource utilization, including CPU, memory, and network bandwidth.
- Fine-tune the application's performance for optimal user experience.

Milestone 9: Monitoring and alerting

Refer to the following:
- Implement proactive monitoring using a combination of metrics, logs, and traces.
- Set up alerts for critical events and potential issues.

Milestone 10: Documentation and knowledge sharing

Refer to the following:
- Document the architecture, deployment process, and best practices.
- Share the knowledge gained through the case study with the organization.

Conclusion

Completing this case study will not only solidify your understanding of multi-cloud Kubernetes deployment but also provide you with a practical roadmap for deploying highly available applications across cloud platforms. By navigating through the project milestones, you will gain hands-on experience in designing, implementing, and managing multi-cloud solutions that meet modern-day demands for availability, scalability, and reliability. Congratulations on reaching the end of this journey and acquiring the skills to excel in the complex world of multi-cloud Kubernetes!

Index

A

access control 121
alerting 255
AlertManager 189
alerts 200
Amazon Elastic Kubernetes Service (EKS) 57, 228
application libraries 106
architecture and design considerations, stateless applications
　configuration management 77
　containerization and image design 76
　data persistence and management 76
　fault tolerance and health checks 76
　interoperability and portability 77
　Kubernetes resources 77, 78
　load balancing and networking 76
　network 78, 79
　scalability, designing for 76
ArgoCD 25, 163, 175
availability 243
AWS Elastic Load Balancer (ELB) 88
Azure Kubernetes Service (AKS) 228

C

client-side libraries 106
cloud-native solutions 255
Cluster Autoscaler 82
comprehensive observability 206, 207
　achieving 211
　best practices 216, 217, 218
　case studies 220, 221
　challenges 219, 220
　implementing 213, 214
　importance, in multi-cloud Kubernetes 207

tools and techniques 211-213
ConfigMaps 64
containerization 2
Container Network Interface (CNI) 9, 12
cost management and optimization strategies 259
cross-cloud communication
  challenges 252, 253
  demonstrating 253, 254
  implementing 252
  secure networking, configuring between AWS and GCP 253
custom resource definitions (CRDs) 15

D

data plane vendor comparison matrix 111-114
data sovereignty 259
disaster recovery strategies 256
Distributed Denial of Service (DDoS) 127

E

Elastic Stack 212
ELK stack 13
Envoy 114
  as data plane 115
  cons 115
  pros 115
  working with, Istio 116
externally generated client requests
  securing 232, 233

F

factors influencing design, HA
  latency 245
  redundancy 245
  replication 245

Fluent Bit 189, 212
Fluentd 189, 212
Flux 25, 163
FluxCD 174
four Cs, of cloud-native security 225
  Cloud layer 229, 230
  Cluster layer 228, 229
  Code layer 225, 226
  Container layer 227, 228

G

GCP GKE
  advantages 250
  features 250
GitOps 25, 155
  automated synchronization 157, 158
  benefits 160, 161
  best practices 166, 167
  challenges 161, 162
  core principles 155
  declarative infrastructure 155, 156
  immutable infrastructure and image-based deployments 158, 159
  operational procedures through Pull Requests 159, 160
  tools 163, 164
  version control, as single source of truth 156, 157
  workflow 162, 163
  workflow, securing 165, 166
GitOps for policy deployment, core principles 171
  automated policy enforcement 172
  automated validation and testing 172
  continuous monitoring and reconciliation 172
  observability and auditability 173

Policy as code 172
security and compliance 173
GitOps implementation, in service mesh 164
  multi-cloud Kubernetes clusters 165
  multi-region clusters 165
  single cluster 164
GitOps policy deployment 170
  implementing 173, 174
  securing 175, 176
  tools 174, 175
GitOps policy deployment best practices 176
  automate everything 178
  clearly defined policies 177
  observability 181, 182
  plan for failure 182, 183
  regularly reviewing and updating policies 180
  testing and validation 180, 181
  version control 179
Google Kubernetes Engine (GKE) 57, 228
  with Istio and Envoy 252
Grafana 164, 189, 211, 255
Grafana Loki 13

H

HA applications 242
  best practices 256-258
  challenges 258
  considerations 258
  definition 242
  importance 242
  Kubernetes and multi-cloud 244
  monitoring 254
  proactive monitoring 254
  service mesh, with Istio and Envoy 244
  unlocking, with Kubernetes 243
  vendor lock-in and interoperability challenges 259, 260
HA applications deployment, on AWS EKS 247
  benefits 248
  EKS cluster, setting up 248
  EKS control plane creation 248
  features 247
  node setup 248
HA applications deployment, on GCP GKE 250
  cluster creation 251
  GKE cluster, configuring 250
HA applications deployment, using GCP-native tools 251
  cloud build and cloud source repositories, using 251
  cloud deployment manager 252
  container registry 251
HA applications deployment, using Helm and other EKS tools
  EKS tools and utilities 249
  EKS, with Istio and Envoy 249
  Helm, setting up 249
HA applications design 245
  factors influencing design 245
  network design 247
  persistent storage 246
  stateful, versus stateless applications 246
Helm 163, 175
Horizontal Pod Autoscalers (HPA) 78
Horizontal Pod Autoscaling 82
hostname 118

hypothetical case study, multi-cloud Kubernetes
   Global HealthTech 32-34
hypothetical use cases, multi-cloud Kubernetes
   compliance and regulatory requirements 32
   cost optimization and flexibility 31
   disaster recovery and business continuity 31, 32
   high availability and redundancy 29, 30
   improved geographical coverage and latency reduction 30

I

Identity and Access Management (IAM) 11
Infrastructure as Code (IaC) 25
infrastructure security
   best practices 230, 231
intra-cluster app-to-app communication
   securing 233, 234
Istio 114
   as control plane 114
   cons 115
   pros 114
   working with, Envoy 116
Istio Metrics 189

J

Jaeger 189, 212

K

Kubernetes 2, 23, 163
   Deployment 96
   Pod 96
   ReplicaSet 96

Kubernetes cluster
   backup and disaster recovery strategies, defining 15
   CI/CD pipelines, setting up 16, 17
   cluster objectives, defining 12, 13
   defining 12
   local development environment, setting up 18, 19
   monitoring and logging, establishing 13, 14
   training and skill development, investing in 17, 18
Kubernetes ConfigMap 64, 94
Kubernetes Namespace 63, 94
Kubernetes Operations (kops) 25
Kubernetes resources, stateless applications 80, 81
Kubernetes security 224, 225
   best practices 236, 237
   case studies 235, 236
   challenges 237, 238
   in context of service mesh 239
   tools and techniques 234, 235
Kubernetes StatefulSet 65
Kubernetes StorageClass 65
Kustomize 25

L

logging 256
logs 200

M

Mean Time to Recovery (MTTR) 215
metrics 199
multi-cloud HA application
   deployment case study 260- 262

multi-cloud Kubernetes
  advantages 25-27
  considerations 28, 29
  disadvantages 27, 28
  evolution 23-25
  hypothetical case study 32-34
  hypothetical use cases 29
  industry best practices 34, 35
  problems 22, 23
multi-cloud Kubernetes environment
  cloud providers, selecting 7
  existing infrastructure, assessing 6
  multi-cloud strategy, developing 8, 9
  need for 2
  network requirements, addressing 9
  objectives, defining 2, 3
  requirements, defining 3-5
  scope, defining 5, 6
multi-cloud NGINX web server system design 87
  architecture overview 87
  load balancing 88, 89
  network design 88
  scaling 89
multi-cloud Redis system design 58
  architectural diagram 62
  data replication and synchronization 59
  implementation steps 63
  key deliverables 60
  networking 59, 60
  project name 60
  project objectives 60
  project plan 60
  project scope 60
  projects stakeholders and roles 61
  project timeline 61

resources 61
risk management plan 62
storage options 59
mutual TLS (mTLS) 110, 132, 142
  implementing, in service mesh 142, 143

N

Network-Attached Storage (NAS) 51
network requirements, multi-cloud Kubernetes environment
  DNS configuration 10
  inter-cluster networking 9
  intra-cluster networking 9
  IP address management 10
  load balancing 9
  network monitoring and visibility 10
  network resilience and high availability 10
  network security 10
  security and compliance, planning for 10, 11
network security 231, 232
NGINX web server 87
NGINX web server deployment, on EKS
  architectural diagram 92
  base configuration 94
  exposing 99, 100
  implementation steps 93
  key deliverables, project 90
  prerequisites 93
  project objectives 90
  project plan 89, 90, 91
  project scope 90
  project stakeholders and roles 90
  public endpoint, testing out 100
  resources 91

resources, creating 97, 98
risk management plan 91, 92
testing 99
verifying 98
versions 93

O

observability in multi-cloud Kubernetes
  complexity 207
  cost management 209
  insights 208, 209
  logs 210
  metrics 210
  performance optimization 208
  pillars 210
  proactive problem detection 208
  traces 210
  troubleshooting and root cause analysis 208
  variability 207
observability, in service mesh 128, 129
observability, with service mesh 215, 216
Open Policy Agent (OPA) 173, 175
OpenTelemetry 213

P

Persistent Volume Claims (PVC) 42
Persistent Volumes (PV) 51
PodSecurityPolicy (PSP) 52
Policy as Code (PaC) 171
policy enforcement 121
pool of servers 126
proactive monitoring 186
  benefits 187, 188
  best practices 200, 201
  challenges 188
  in service mesh 202, 203
  monitoring stack, securing 201, 202
  tools and techniques 189, 190
proactive monitoring implementation
  alerts, setting up 195, 196
  dashboards, building 194, 195
  iterative improvement 197, 198
  monitoring setup, testing 196, 197
  system, understanding 190, 191
  tools, configuring 193, 194
  tools, selecting 191-193
Prometheus 164, 189, 211, 255

R

Redis deployment
  base configuration 63, 64
  exposing 69, 70
  prerequisites 63
  public endpoint, testing out 70
  resources, creating 67
  StatefulSet and PVC 65
  storage class 65
  testing 68
  verifying 68
  versions 63
reliability 243
resiliency 122
resiliency implementation 122, 123
  circuit breakers 124, 125
  health checks 125, 126
  load balancing 126, 127
  outlier detection 128
  rate limiting 127, 128
  retries 123
  timeouts 124
Retries 123

role-based access control (RBAC) 11

S

scalability 243
Secure Sockets Layer (SSL) 137
security policies, in service mesh
    control plane 145, 146
  audit logging policies 147
  authorization policies 146
  network policies 148
  peer authentication policies 148, 149
Service Level Objectives (SLO) 186
service mesh 104, 105, 213
  control plane manager, usage criteria
    110, 111
  data plane manager, usage criteria 112,
    113
  history 106, 107
  implementing 108, 109
  observability 128, 129
  operational benefits 105, 106
  use cases 107, 108
service mesh components 109
  control plane 109, 110
  data plane 110
service mesh security 132, 133
  access control 143, 144
  authorization 143
  considerations 150, 151
  isolation 144, 145
  network policy 144
  traffic encryption, with mutual TLS 142,
    143
service mesh security principles
  auditability 141, 142
  fine-grained access control 137, 138
  least privilege 135

  mutual authentication 136
  observability 139, 140
  policy enforcement 138, 139
  secure communication 136, 137
  Secure Ingress/Egress 140, 141
  zero-trust network 133-135
service mesh traffic control 116, 117
  fault injection 119
  load balancing 117, 118
  quota management 120
  rate limiting 120
  routing 117
  service discovery 118, 119
  traffic mirroring 119, 120
shadowing 119
Site Reliability Engineering (SRE) 186
stateful and stateless application-based
    clusters, differences
  data persistence and management 39
  deployment and maintenance
    complexity 40
  high availability and fault tolerance 39
  scalability 39
stateful application-based clusters 41
  considerations 51, 52
  Kubernetes and stateful applications 42
  multi-cloud Kubernetes deployments 42
  stateful applications 41
  use cases 43, 44
stateful application deployment example
    57
  challenges, for Redis deployment 57, 58
  multi-cloud Redis system design 58
stateful application Kubernetes cluster
    design patterns 56
  active-active pattern 56
  active-standby pattern 56

backup and recovery pattern 57
caching pattern 57
leader election pattern 57
multi-cloud pattern 57
persistent storage pattern 56
replication pattern 56
statefulSet pattern 56
stateful applications 38
  best practices 55, 56
  challenges 51
  data storage strategies, across multiple cloud providers 52, 53
  overview 50
  real-time data across clusters, sharing in different cloud providers 54
StatefulSets 42
stateless application-based clusters 44
  advantages 47
  challenges 48
  complexities 45, 46
  implications 45
  use cases 46, 47
stateless application deployment patterns 82
  blue-green deployment 83
  canary deployment 83
  multi-cloud deployment pattern 83, 84
  single cloud deployment pattern 83
stateless applications 38
  advantages 75
  architecture and design considerations 76
  auto healing 82
  challenges 75
  Cluster Autoscaling 82
  examples 74
  Horizontal Pod Autoscaling 82
  Kubernetes resources 80, 81
  load balancing 81
  overview 74
  pod scaling 81
stateless applications system design and project plan 84, 85
  challenges, in deployment 86, 87
  example application 85, 86

T

Terraform 163, 175
threat modelling 149
  security best practices 149, 150
traces 199
Transport Layer Security (TLS) 122, 136
two-way authentication 136

V

Vertical Pod Autoscaler (VPA) 89
virtual machines (VMs) 18
Virtual Private Cloud (VPC) 10
Virtual Private Networks (VPNs) 28

Z

Zipkin 189, 212